How to Save the City

How to Save the City

A Guide for Emergency Action

Paul Chatterton

Illustrated by James McKay

agenda
publishing

*Dedicated to everyone, past, present
and future, saving the city – our breakaway
coalition of emergency first responders*

First published in 2023 by Agenda Publishing
Reprinted 2024

Agenda Publishing Limited
PO Box 185
Newcastle upon Tyne
NE20 2DH

www.agendapub.com

ISBN 978-1-78821-478-0

British Library Cataloguing-in-Publication Data
A catalogue record for this book is available from the British Library

Typeset by Patty Rennie

Printed and bound in the UK by 4edge

Contents

Acknowledgements

To produce a book like this there are so many people to thank. First, projects and groups along the way have provided opportunities for me to learn, gather and try out ideas in this book – Lilac Cohousing, Kirkstall Valley Development Trust, Leeds Community Homes, Our Future Leeds, Climate Action Leeds, the Leeds Doughnut Coalition, Leeds Love it Share it, the Racial Justice Network, the Trapese Popular Education Collective, the Camps for Climate Action, Radical Routes, the Doughnut Economics Action Lab.

Many people have provided support and enthusiasm – my wonderful friends at Lilac Grove, colleagues in the School of Geography, and fellow climate activists at Our Future Leeds and Climate Action Leeds – Andy, Simon, Margo, Jenni, Andrew, Matt, David, Jefim, Maia, Tim, Helen, Irena, Anzir; and as always, my partner Tash and sons Rafi and Milo for being my guides and inspiration.

Special thanks to James McKay for providing the captivating illustrations that bring this book to life, Stella Darby for her support with referencing, editing and ideas, and Jenny Bull for designing my illustrations. Finally, thanks to the team at Agenda Publishing and especially Steven Gerrard and Camilla Erskine for knowing that a book like this is urgently needed.

Paul Chatterton
Leeds

1

Introduction

In 2018, dozens of people died in Tokyo as temperatures exceeded 40°C for several continuous days, and in California the town of Paradise was largely destroyed by a wild fire in just a few hours killing 85 people. In the summer of 2021, deadly flash floods killed scores of people in the Bad Neuenahr-Ahrweiler area of Germany. In early 2022 Cyclone Batsirai left 120 people dead and nine out of ten homes destroyed in Madagascar's coastal city Mananjary, while later that year residents of Jacobabad in Pakistan endured temperatures almost intolerable to humans at 51°C.

As these climate disasters unfold, our urban world continues to grow. One recent estimate suggests that 290,000 km2 of natural habitat could be converted to urban land between 2000 and 2030, with huge implications for habitat and biodiversity loss (McDonald *et al.* 2020). In the UK, between 2006 and 2012, 22,000 hectares of green space has already been lost to urban sprawl, the equivalent of an entire city, while in Shanghai urban greenspace declined by 30 per cent between 1980 and 2015 (Wu *et al.* 2019).

Equally worrying are the persistent social challenges facing our urban world. In Africa only 15 per cent of residents in Lagos and Kampala have piped water to their dwelling (Beard & Mitlin 2021), while in the Democratic Republic of Congo and Chad only around 40 per cent of the urban population have access to electricity (Our World in Data 2020). A recent study found that 51,000 premature deaths could be prevented each year across 1,000 European cities if they achieved levels of small particulate air pollution (PM2.5) recommended by the World Health Organisation

(Knomenko *et al.* 2021). In New York, according to the Eviction Lab (2022), there were almost 180,000 evictions filed by landlords between 2020 and 2022.

Getting into emergency mode . . .

These examples are not isolated or trivial. What we make of them is the core argument of this book. They are a selection of events and facts that represent the broader climate, ecological and social emergencies currently experienced in our urban world. We now know that our global climate system is in breakdown and that this has unquestionably been caused by human activity over the last couple of hundred years since the industrial revolution and fossil-fuel capitalism; we know that scientists have warned that unless there are immediate, rapid and large-scale reductions in greenhouse gas emissions stopping dangerous levels of global heating will be beyond reach; we know that the fast-growing model of urban development based on ceaseless economic growth is pushing our natural world to its limits, with almost half of non-human species now under threat; and we know that the long-standing social challenges of poverty, inequality, hunger, malnutrition, oppression and fear are everyday realities for too many people. We know that these converging emergencies are not natural phenomena, but human-made problems, lock stepped with the growth of our globalized, corporate-dominated, industrialized urban world.

How to Save the City is a wake-up call about what we can all urgently do faced with what I call the accelerating and converging triple emergencies – climate, nature, social. Transforming our cities is no longer discretionary. Without decisive action we are facing the possibility of an Earth that is not safe and habitable for humans. In stark terms, there are no cities on a dead planet. This book is a guide for what we can all do to mount emergency responses, in order to, literally, save the city. We have a huge amount to do in the decade ahead; every year counts and immediate changes are required. Our efforts need to focus on all three emergencies. A core focus of this book is that this is not just about reducing greenhouse gases, although that is a central priority. As we create post-carbon city futures, we must also tackle the longer term ecological and social emergencies. You might be a planner changing local regulations, a teacher looking for course material, a student or researcher writing an assignment, a politician leading change, a city official lobbying for a new policy, an entrepreneur supporting local well-being or an activist developing campaign skills. Whatever your role in city life, this book can support you to get into emergency mode, and galvanize your networks, communities, workplaces and institutions to this task.

. . . as emergency first responders

As a guide for emergency action, this book has a particular approach. We know that the idea of emergency can instil panic, fear and a sense of powerlessness. But this is only one set of responses in the face of emergency. We panic when we do not have the skills and experience to deal with an issue. An improved understanding of the challenges ahead and the kinds of responses that work can support our ability to act and think clearly about the options available to us, and what needs to be done. Therefore, rather than seeing ourselves as passive recipients of terrifying and overwhelming emergencies, I encourage us to see ourselves as first responders on the scene tackling the climate, nature and social emergencies. A first responder is someone who has a basic level of training and is the first on the scene of an emergency. They are part of a longer tradition of non-professional civil society action, community-based volunteering and mutual self-help work where ordinary people provide services and assistance often during emergency and war-time situations.

Seen in this way, we can avoid the barriers of just relying on professionals and experts, and explore how being an emergency first responder can fit into our own lives, experiences and skills. The broader intention is not to enter into a drawn out and exhausting emergency mode, where we are on constant red alert against each other and an invisible enemy. Emergency situations can offer an opportunity to get equipped with the insights, skills and connections to make a difference and act effectively, rather than freeze through despair, denial or anger. Our emergency action needs to be urgent and decisive yet also empowering and inclusive. It needs front loading with a burst of action that creates immediate gains for climate, nature and people every year from now, laying down the conditions for a better urban world as we go through our decade of transformation into the 2030s.

While we cannot fully control what is ahead, we are all capable of empowering and skilling ourselves and others to be able to understand the challenges, and what responses and solutions we can develop now to build a safer city. This is not the work of individuals working in isolation. To effectively respond to the challenges ahead, we need what I call a "breakaway coalition" of emergency first responders – a network of people across business, research, civic and public sectors who are prepared to support and build a new kind of city based on equality, regeneration and the common good rather than extraction, ceaseless economic growth and hierarchy. It is these first responders who can deploy the guide for emergency action I outline in this book.

What I propose in the coming pages is not some utopian fantasy. It is based on 20 years of action-based academic research on workable solutions that can meet the converging emergencies and set a course for a safer future. How things will turn out is impossible to predict. Therefore, this book is also based on hope and imagination, inspired by examples of positive practical action that actually work in the here and now. With clarity on the challenges, strategic thinking and collective action, a coalition of actors can indeed save the city, and change it for the better.

City politicians and campaigners are waking up to these emergencies. Billionaire and former Mayor of New York Michael Bloomberg boldly stated that cities across the world should mount a challenge to the existential threat of climate change (Bloomberg & Pope 2017). In 2018 London became one of the first cities around the world to declare a climate emergency, in an effort, as London mayor Sadiq Khan put it, to meet the existential challenge of climate change (Taylor 2018). Since then, thousands of localities around the world have followed this lead and declared climate and ecological emergencies. A year later the protest group Extinction Rebellion held its International Day of Rebellion with thousands of protesters engaging in acts of non-violent civil disobedience in 60 cities around the world, while the Global Youth Strike, inspired by Greta Thunberg, mobilized millions of young people across the world to strike for climate action.

Grassroots networks for change such as the Climate Emergency Alliance, 350.org, Build Back Better, Green New Deal Rising, the Rapid Transition Alliance, Fridays for the Future, Black Lives Matter, Decolonising Economics, and the Doughnut Economics Action Lab represent a new zeitgeist of ideas and skills for our emergency action. They share an approach that is key to this book. We need to mount responses that regenerate nature and mitigate against climate breakdown in ways that also tackle long-standing urban problems – poverty, alienation, segregation, social isolation, racism, violence, precarious work, corporate greed and powerlessness. Whoever we are, and whatever we do, we also have to realize that change will not come easy. Our efforts to build creative alternatives and new exciting coalitions will need to be paired with resistance and direct action if we are to disrupt powerful interests and create the urgent, bold and decisive transformations required. As the triple emergencies converge and amplify, we will be forced, largely out of necessity, to use more drastic measures to ensure a safe future for all.

About the book

How to Save the City is a provocative and disruptive, yet playful and game-inspired, book. I organize it around three main chapters – strategy, players and moves – that together represent a systematic way of approaching emergency change in your place. This structure provides the ideas and practical tools needed for you to move through the city as change-makers in an emergency context. Let's take a look at these three aspects.

Strategy: our approach to change

First, we need to deal with strategy. A strategic approach creates a plan of action designed in a way to achieve an overall aim. Starting with strategy encourages us to think about how change happens, what some call a theory of change. Having a theory of change is nothing new and indeed not unique to grassroots, radical or civic movements. It is used by groups across the political spectrum who want to intervene in how change happens (van Tulder & Keen 2018; Alexander & Gleeson 2020). This part of the book, then, outlines the strategic rules of the city game, and how you can use them to make emergency change.

I base my strategy around what I call the Learn–Act–Build approach drawing on a wealth of thinkers, activists and campaigners that have worked on change over the years. These include Joanna Macy and Chris Johnstone's work on "the Great Turning" in their book *Active Hope* (2012), Eric Olin Wright's *Envisioning Real Utopias* (2010), Kate Raworth's *Doughnut Economics* (2017) and Bill Moyer's *Movement Action Plan* (2001).

Learn, act and build do not all have to happen at the same time. But they have to feature at some point in what we do. First, we have to learn new ways of knowing and being in the world and unlearn those that harm and stop change; second, we have to act, resist and make a stand against that which causes damage; and third, we have to build alternatives that make real the world we want to see.

Players: who will do it?
Second, once we have a strategy, we need to explore who will use it and what they can do to get into emergency mode. These are the players – groups and individuals we will meet and engage with in our quest to save the city. I have selected nine players. Clearly, this is not an exhaustive list, and they will vary immensely across social contexts. Consider them "pen portraits" to stimulate thinking and action about the breadth and diversity of actors involved in emergency action. We need to know how these players work, what role they can play in saving the city and what social power and resources they can leverage. Just as interesting is how we can work between and beyond these personas. How can we integrate amongst them – sharing lessons and making coalitions that can break away from the current status quo. One of the real opportunities is how we can operate outside some of the narrow institutional roles in the world of work and politics that maintain things the way they are. We need to recognize that we all have multiple roles to play, and skills to offer, in reclaiming a safe future.

Each player has a particular role to play as they shift into emergency mode. These include:

- The scientist/researcher and how they can create an experimental approach to the city where our places become living laboratories for emergency solutions to the problems we face;
- The teacher/educator and how we make use of teaching and education to relearn the art and skills of community so we can navigate the emergencies ahead;

- The entrepreneur and how we build a new economy which has at its heart the communities it serves and embeds city wealth building;
- The city-maker and the skills of being an interconnected practitioner that can work across silos and departments and lead a breakaway coalition;
- The social activist and how everyday activism can harness broad social power in ways that involve and engage people to take action;
- The consumer and how we link buying and making, instilling an ethics of care into our daily trading which becomes more local and community responsive;
- The citizen and how we manage the urban commons, bringing the city's land and buildings into use for communities to build resilience and sustainability;
- Big business and how we harness their power and skills as a force for transformation through a circular and regenerative economy;
- The non-human and how humans can reconnect with plants and animals and rewild the city in the face of the rapid extinction of the natural world.

Moves: making it happen

Finally, what will our players do with their newly acquired strategic approach? Here, I present emergency action areas, what I call moves, that our players can do to save the city. I offer moves in areas that will resonate with modern city life, and importantly underpin thriving, safe and equitable city futures. Each move topic is essential to tackling the emergencies we face and I have framed them in a way that highlights multiple gains across climate, social and nature challenges. I want us to think about the inter-connections between these move topics – how change in one area impacts another. As we explore these moves, the key aspect to interrogate is the roles of the different players and how our strategic approach informs and guides them. More than anything, I want us to see these move topics not as static things to change in isolation, but as broader city systems that combine to create transformative system change and an urgently needed basis for living well within the limits of our natural world.

Emergency move topics include:

- Mobility. This is the big one. Fossil fuel vehicles are accelerating city carbon emissions and a whole basket of problems around air quality, road deaths and nature loss. I call this move "car-ism": a largely invisible culture of the car that has taken over our cities from top to bottom. I explore how we can unravel car-ism to unlock a very different mobility system that is affordable, efficient and a route out of our emergencies.
- The economy. Most of the challenges we face can be traced back to the economy. What we do, make and buy impacts us, others, and the natural world. I explore how we can build a new city economy based on community-responsive businesses and workplace democracy that can allow us all to thrive within the limits of our planet.
- Placemaking. Planners, architects and developers make places to live. But this move is about a very different agenda – what I call climate emergency placemaking. How can we shape beautiful, affordable places while also adapting to climate change, reducing dangerous greenhouse gases and restoring nature?
- Aviation. This is one of the few areas of city carbon emissions without a plan. The aviation industry advocates constant growth while claiming it can tackle the climate emergency based on untested technologies and sustainable fuels. Without slowing aviation growth and rapidly embedding alternatives, this is one area that will compromise our shared safe future.
- Democracy. City Halls and local decision-making often feel distant and out of touch. We need to renew local democracy, participation and engagement to unleash the power of communities to tackle the triple emergencies.
- Nature. As human demand for space continues to expand, the

extinction of non-human life on this planet continues at a rapid pace. How we create a new deal for nature and reverse the mass extinction of our fellow species is one of the key challenges ahead. This is not a "nice to have" addition. Our ability to live safely depends on the survival of other species.

It is best to read this game-inspired book sequentially. Strategies lay a bedrock of ideas that help illuminate our understanding of the role of different players, which then informs our moves. But equally, rules are there to be broken. That is the point of saving the city. We will have to break out of existing ways of thinking and acting. Clearly, no book can reflect the diversity and complexity of our urban world. There will be much I have left out, or places where I have not captured nuances. Additions and amendments need to be teased out in further conversations that I hope this book prompts.

This book, then, represents an unfinished and ongoing story. It is about what might happen over the next decade. For this reason, the final chapter is a short description of life in the early 2030s around the ideas in this book. Imagine if they all happened – if we actually saved the city, retrofitted and shaped it along the lines in this book. The next decade is also a story of people who will be born into, or come of age in cities that they did not directly make or shape. But they will take up the story. My account is told through the voice of a teenager – reflecting the many young people, family, friends and students I am surrounded by in my own life. Stronger bonds of allyship and support will be required between the adult world and young people to ensure the needs of future generations are at the heart of saving the city. The book also features a collection of cartoons by graphic illustrator and novelist James McKay. These cartoons feature three young people in dialogue as they explore what saving the city means for them.

The following chapters feature specific ideas for emergency action, some current examples to check out (I have purposefully not provided weblinks as these change), and a list of web resources at the end. These ideas, examples and resources can be used to create better city futures right now. I have chosen them as they push beyond what is already possible, familiar and known. They are stretch activities that rise to the challenge of our triple emergencies and take us beyond the business-as-usual model of urban development. But I also acknowledge the complexities and structural barriers that still exist to these kinds of ideas – especially in terms of the lack of political power, financial resources, and broader trends such as rightward shifts in politics, increasing corporate control, war, cost of living crises and austerity. Clearly, there are also versions where the future did not turn out all right, where urban life continues to degrade or even collapse, wreaking havoc and misery. Those scenarios are for a different book, but tragically they are also all too real for many people around the world faced with conflict, oppression, forced migration and violence. For now, I am convinced that we can save the city for the better. The window of opportunity is small and reducing. But as emergency first responders we have all the ideas, solutions and energy to take decisive and urgent action.

How do you save the city?

Over half the world's population is now urbanized. This urban world is dependent on dangerous levels of dirty fossil fuel energy, is responsible for about three-quarters of global greenhouse gases, has ecological footprints far larger than their actual size, remains locked in to high-growth, high-consumption lifestyles, and continues to generate inequality and precarious living conditions for many. In this deeply urbanized world with all its attendant problems, to save the city feels like a broader quest to save humanity. The idea behind this book is to ask a very direct question: in a time of converging emergencies how do you save the city? And what is this entity – the city – we are trying to save, does it need saving, who will save it, and what does saving mean anyway? Is there something obvious we have not found or done?

What is a city?

While I use the city as a frame for this book, it is short-hand for many different experiences. The ancient preindustrial world created walled citadels across Europe, Asia and Latin America. The nineteenth century gave us the industrial city of smoke, squalor and poverty, while the twentieth century created the modern car-based city with its suburbs and retail malls. For some, the city equates to downtown pedestrianized areas of cafes and shops, glazed offices and high-rise apartments, and close proximity to others. More recently, urban areas have expanded which hardly resemble these city forerunners. Two billion people live in informal settlements (negatively known as slums) in exploding megacities in the global south, while many live in complex polycentric urban areas, smaller towns, edge suburbs or peri-urban environments (Satterthwaite *et al.* 2020). The contemporary planetary urban experience, then, diverges greatly from the traditional city ideal of a central district with surrounding areas set in neat concentric circles. As the social ecologist Murray Bookchin (Bookchin & Foreman 1991) reminded us, what characterizes the present age is a process of urbanization that is creating human settlements that few would recognize as traditional cities.

While the majority global experience is now an urban one (Brenner 2015), urban land only makes up a fraction of the world's surface. But urban space has impacts way beyond its physical limits. On one level cities are bounded entities with edges made up of a collection of material things – buildings, pipes, technologies, wires, microchips, consumer goods, bodies, animals and plants. Beyond these real things, cities are made from flows and networks of information, goods, power and money that spread far and wide. It is often the flows that define our lived experiences more than the material things. Urban theorists use the idea of assemblage to refer to the combination of actors, human and non-human, place based and external, that interact to create city life (Farías & Bender 2011). As I explore in the next chapter, our strategy needs to be alive to this complexity in our quest to change the city.

Amongst the triple emergencies that motivate this book, cities are seen

as both saviours and villains. Balanced between metrophilia and metro-
phobia (Waite & Morgan 2019), cities are loved and loathed in equal
amounts. There is a long tradition of regarding the city as a pathology,
focusing on public health, crime or disorder. It often seems like cities
by their very nature are simply bad for our health. Cities are the source
of many of the world's modern ills, and perhaps they can never be truly
sustainable. They have deep roots leading back to colonialism, conquest,
extractionism, hierarchy and oppression. The logic goes that once humans
started to settle thousands of years ago and build what we now know as
civilization, they laid down the seeds of their own demise. So while modern cities have a veneer of civility, ultimately they are predicated on the
domination and conquest of nature and the exploitation of the powerless many by the powerful few (Jensen 2006). At the same time, many
urbanists celebrate them as crucibles of innovation and creativity, concen-
trating talent, wealth and resources that power the prosperity of nation
states (Glaeser & Cutler 2021; Florida 2004). Clearly, we need to critically
explore this approach in terms of who controls and benefits from innovation and prosperity. As I explore later, the more exciting potential is to see
the city as home to the multitude, where shared existence and encounter
between people can open up possibilities for making a new city commons
(Hardt & Negri 2004).

The reality of what a city is, therefore, is complex (Barnett 2016). There
is nothing instinctively good or bad about cities; it is how city life is organized, governed and owned that is crucial. We need to enhance tendencies
that support thriving lives and restrict those that do not. We also have to
start from cities as they exist now, and use these as resources to rebuild
and retrofit the future we need. As emergencies collide and amplify, cities and their residents have the possibility to rapidly try out and embed
solutions. One reality that will not go away is that in an age of existential crisis, cities, and especially their higher consuming, higher resource
using residents and workers, carry a great responsibility for the survival
of humanity in the twenty-first century. As more and more of the world's
population live in cities, or at least sprawling urban areas, we need hope

that this increasingly urban world will sow the seeds of renewal, rather than downfall. We have to unlock the potential of cities to navigate us through the next years.

It is likely that what we know as the city may become unrecognisable. Emerging human settlements may need to embrace novel forms and functions that bear very little resemblance to past versions in order to tackle future crises. As we emerge from the Covid-19 pandemic, the kinds of settlements that underpin flourishing lives will remain hotly debated. This book is a guide to urgently accelerate trends that will be beneficial going forward, and to slow others that are less so. What will take centre stage is the need to rein in the power of the world's super rich and mega corporations and redeploy their vast resources towards a climate safe, equitable and greener urban future.

What does saving mean?

Does the city even need saving? Is the city something that needs to be controlled and corralled, or is the task to stand back and set the conditions for its creative potential to flourish on its own terms? Childhood psychologist Alison Gopnik (2016) asks are you a carpenter or gardener? The same can be asked of city activists, planners, researchers, politicians and entrepreneurs. Do you see your task as sculpting city life to a set vision and outcome, or creating protected but diverse spaces for experimentation, learning and play where solutions and natural resilience can grow? Can city elements be somehow co-ordinated for the greater good? Is co-ordination even possible, to make the modern metropolis hum like a football team or orchestra? How do we sum the efforts across millions of people?

Many might say we simply need to get better at what we are already doing – better urban management, more trickle down from economic growth, more control of complexity. But given what we know about the shortcomings of these approaches, we need fundamental system change that can embark on a wholescale rethink of how we design and live in cities. A key part of the response will be the role of new technologies. Much emphasis is being placed on the "smart" city – the idea that aspects of city life across energy, transport, food and housing are mediated by advanced automated information and communication technologies. Clearly, this calls into view the huge current gaps in access to digitally-enabled city life. A handful of large global corporations – IBM, Microsoft, Cisco – offer cities and communities a range of smart solutions that are beginning to run and manage large aspects of our lives (Hollands 2008). Transport ticketing, waste management, home heating, street surveillance, parking permits, heat sensors, and social media feeds are joined together in what is known as the Internet of Things (IoT) (DeNardis 2020). There are benefits here. Real-time data in city dashboards can identify and respond to problems. The key question is in an age where the internet penetrates more areas of our life, who owns this data and to what use is it deployed? Is this the beginning of a city of total surveillance, or our liberation through advanced and artificial technology?

Even more crucial, then, is how we manage and own technologies. Given our technology dependent society, a tendency is to assume that another technology will arrive to save the day – something that will perhaps liberate us from fossil fuel dependency, carbon emissions or low-paid precarious work. Clearly, technological innovations will continue. We will need more of these, especially in areas such as energy storage and generation, micro-mobility and sustainable construction, but less so in areas such as media, finance and militarization for example, where tangible public benefits are less clear. The key challenge for us going forward, is how to guide technological innovation to ensure the benefits are commonly controlled and broadly socially useful (Bridle 2018).

What if we cannot save the city, or parts of it? What if we have passed some threshold beyond which urban life is salvageable and is degenerating rapidly beyond our control? How do we learn to live with this loss, our inability to save or recover what we cherish and the sadness that goes with that? This is certainly the case in many parts of the world already subject to conflict, poverty and pollution. We can relate to this through lost local greenspace, now disappeared under the forward march of city development. More controversially, what if cities, or at least parts of them, are not worth saving; that we need to accelerate their demise to make way for a safer and inclusive form of human settlement? What elements of city life are worth maintaining and which should we jettison for the common good? What elements can we repurpose and retrofit, or metrofit as Tony Fry (2017) suggests. What do we do if solutions head towards greater segregation, social control, walled ghettos and restrictions on movement? These issues need to be tackled head on.

This brings us to a further complexity. What does success, or indeed failure, actually look like? Is there a moment that we can claim victory, and who do we claim it over? There is no global referee who can proclaim we are on track. What matters will be judged locally, and probably by future not current generations. One thing is sure, that both successes and failures will be uneven and existing inequalities across the world will continue to be enhanced unless there are purposeful efforts to identify and address them. Many parts of the world are already in a process of slow and enforced decay be it from war, famine, political corruption or lack or resources to invest in resilient infrastructure. Success is also deeply relative. For some, it might mean saving a local park, closing a coal power station, or changing government policy. For others it might simply mean surviving in the face of an encroaching army, police brutality or the long-term effects of poverty or racial and gender violence.

Who will save the city?

Finally, it may provide comfort to think that there might be someone in charge who will save the day. The very idea of heroes and saviours are deeply engrained in the myths of human civilization. But what if they are part of the problem? Part of saving the city means to disrupt gendered and racialized power relations and ensure actively excluded groups get structural power (Kempin Reuter 2019). We urgently need to broaden the ownership of city solutions away from established technocratic and political elites and a narrow range of options. What is especially important is to create more inter-sectional encounters across, race, class, gender, sexuality and physical ability, to distribute structural power, disrupt how existing decisions are made and resources allocated. A big part of saving the city is to take seriously calls for anti-racist, anti-sexist, anti-ableist and anti-adult practices.

Maybe the whole idea of saving is simply too arrogant. Who are we to think we can save the city? Perhaps city inhabitants are their own greatest enemy, unwittingly bringing forward the demise of the city they depend upon in the pursuit of comfort and convenience, turning a blind eye to the great corporate takeover of our urban lives. Or maybe the problem lies far away in board rooms that citizens and local politicians have no control over. What is the point of even trying to mount a response, if decision-making and power are somewhere else, out of sight, out of reach. As I explore in our strategies, the answer is to adopt a complex geographical mindset, where we act within, outside and across the city.

Where does agency and power actually reside? The city is often referred to as if it were its own actor that has agency over our future. It is almost as if there is some higher force at play beyond the people that live there. Can the causes and solutions be found amongst the structures, plazas, buildings and institutions we call the city? Or do we, the inhabitants have the agency to change the immediate future? The answer is likely both. To paraphrase Marx, people make their own cities, but they do not make them under circumstances they choose. These circumstances already exist, and are transmitted from the past. There are always larger historical forces at play, but we can attempt to shape them. For example, we might not abandon the city hall, but we can change the way it functions for the emergencies ahead.

What makes us think the city needs saving *now*? Are we falling into another generational trap that this is the make-or-break moment, the last chance to mount an emergency response? Older generations will say they have been here before, that they lived through moments that felt like the last chance for change, and then life continued. Our converging triple emergencies are different in key respects. We now have stark scientific warnings about what will happen if we do not take rapid, urgent and transformational action. There is an additional element of inter-generational equity. The current younger generation and those to come will deal with an emergency situation not of their making. Understandably they will ask "why did the last generation fail to act?" There is something unique about the convergence of emergencies in this moment. Atmospheric physics does not care what we think. Greenhouse gases will continue to accumulate rapidly in the atmosphere whether we feel and acknowledge the urgency or not. The challenge is to build meaningful policy and solutions around these nature, social and climate warnings which have been growing for years, but are now colliding and accelerating in a way that threatens our very existence. The worst that could happen if we over-estimated the challenge is that we build a better world for nothing.

* * *

Before I turn to the strategy underpinning this book, in the next chapter I explore the scale of the emergency situation we face, why we have to act now, and what our decade of transformation to the 2030s can look like. We have to get real about the severity of climate breakdown, nature loss and social inequality, and use this to get into emergency mode. This is essential knowledge we can all acquire so we can be emergency first responders and play our part in saving the city.

2

Our decade of transformation

Do we really only have ten years left to save the planet? What does this mean for immediate action, and where do we need to be by the 2030s? Given this really is an emergency, what can we all do to empower ourselves, understand what is happening and respond appropriately. Amongst the inaction, denial and confusion, how can we get into an effective position as first responders to our triple emergencies? Let's start by exploring some of the science and policy that needs to guide our emergency action.

In 2015, the Paris Agreement was signed by most of the world's nations to limit global heating to preferably 1.5°C, compared to pre-industrial levels. This legally binding treaty was a watershed, marking a moment of hope that the international community could galvanize emergency action. Very quickly it became apparent that not enough progress or commitments were being made to make the Paris Agreement a reality. Leading scientists continued to put out further dire warnings. In 2018, the United Nations Intergovernmental Panel on Climate Change (IPCC 2018) made a rare bold statement in their special report that avoiding dangerous levels of global warming would now require rapid and far-reaching transitions in energy, land-use, built environment and infrastructure which are unprecedented in terms of scale. This was distilled for policy makers and the media as roughly a decade to halve greenhouse gas emissions on a journey to zero by about 2050, but equally that immediate and urgent action is needed year on year to achieve these.

The 2021 Glasgow climate talks at COP26 reinforced the same dire situation and lack of progress. Without a huge step change in commitments

and action the Earth is heading to be a much hotter and less safe place for humans. In the same year in their 6th Assessment Report the IPCC (2021) stated that unless there are immediate, rapid and large-scale reductions in greenhouse gas (GHG) emissions, limiting heating to 1.5°C will be beyond reach. This matters. One and a half degrees of heating does not sound much in the greater scheme of things. Our world is already about 1°C hotter and as a result the last few years have broken record after record in terms of floods, droughts, mega fires and super storms. This extreme weather is wreaking widespread damage now and is only a taster of what is to come as we throw more heat into the Earth's finely balanced natural systems. With every increment of global warming, especially above 1.5°C, there is a whole basket of consequences – each with greater impacts – on food, energy, housing, disease, migration, biodiversity as well as civil peace. Scientists and ecologists have long understood there are tipping points, where the Earth's systems can act in unpredictable ways through positive feedback loops that spill into other areas. Consider a stone rolling down a hill; once it starts to roll it builds up momentum until it becomes unstoppable and irreversible. Once we get to 1.5°C, the task to reverse global temperature rises becomes much harder. We currently may be on course for a planet that is 3 or 4°C hotter by the end of the twenty-first century. As David Wallace-Wells (2019) frankly puts it, that means an uninhabitable planet.

Scientists have given us a collection of adjectives (rapid, immediate, unprecedented, large-scale, far-reaching) to frame our actions. What do these mean for city leaders, entrepreneurs, activists, researchers and citizens? What can we all actually do differently right now in our daily practices to make the intent of these messages from global scientists real as soon as possible? How can we radically decarbonize places, while also ensuring social justice and nature recovery is at the heart of what we do? How can the allocation of resources, decision making and organization put into action rapid, immediate, unprecedented, large-scale and far-reaching changes? Let's start by exploring the urgent task of city carbon reduction.

Don't blow the carbon budget

Meeting that now-sacred target of 1.5°C heating is all about keeping to a specific budget for the amount of GHGs we produce. When we burn fossil fuels such as coal, oil and gas we emit carbon dioxide (CO_2), and that acts as a greenhouse gas that in turn increases global heating. Greenhouse gases are measured in CO_2 equivalents (CO_2e) as there are other greenhouse gases such as methane, and these are also created from human activity, especially intensive animal farming. We measure CO_2e in our atmosphere in parts per million (ppm). The scientific community suggest that a safe concentration is 350 ppm. We are currently approaching 420 ppm at a rate of over 1 ppm per year. This is not a safe place to be.

The IPCC (2018) stated that to give humanity a two-thirds chance of holding to that 1.5°C target, humanity has an all-time (or to 2050) budget for the emission of GHGs of around 420 gigatonnes (a gigatonne is a billion tonnes, and a billion tonnes is a thousand million tonnes). So even sticking to this budget we cannot guarantee that temperatures will remain at or below the 1.5°C target – especially given that we are currently already over 1°C of warming. With global average increases of 1.5°C, there will still be a less safe world – continued species loss, extreme weather, food shortages and sea-level rise. The latter part of the century will require further action to undo the damage to the climate by humanity over the last few decades, as well as the deployment of as yet unknown technologies, to return to climate safety.

We need to put these huge numbers into perspective. A tonne of carbon dioxide does not sound that much. The average citizen of the USA is responsible for about 16 tonnes a year, while the global average is about four tonnes. A typical passenger vehicle emits nearly five tonnes of carbon dioxide per year. But there are lots of us on the planet, and a small group of now well-developed countries have been burning fossil fuels and generating those global heating gases for a few hundred years now. The UK, closely followed by France, Germany and the USA were the early forerunners here, recently joined by emerging economies such as India, China and Brazil. This is a story of uneven global use and responsibility, but this story of unevenness is also changing. What we now know is that the burden rests on high-income high consumers, who are no longer restricted to the rich global north but spread across many nations of the world, and it is this global super rich in particular who are having the biggest impacts.

From the industrial revolution in the late eighteenth century until now, about 1,500 gigatonnes (billion tonnes) of GHGs were emitted from human activity (Stainforth & Brzezinski 2020). Some of this is reabsorbed by the planet's carbon cycle. But taken over the last few million years, it is an unprecedented amount. The jaw-dropping aspect is that half of these GHGs were released since 1990, the year the IPCC launched its first assessment of the problem (*ibid.*). A key issue is how much more fossil fuel reserves are left in the ground, and how much of this is what is called "unburnable carbon". While there are some controversies about figures, one estimate suggests there is up to 11,000 gigatonnes of remaining fossil fuel reserves (Jakob & Hilaire 2015). A further study used the idea of carbon bombs (oil and gas projects that have over 1 gigatonne of CO_2 potential) to identify 425 such projects that combined have a CO_2 potential of over 1,000 gigatonnes (Kuhne *et al.* 2022). To stay within that safe 1.5°C limit most of these reserves are unextractable. Recent research, for example, suggests that "nearly 60 per cent of oil and gas, and 90 per cent of coal cannot be extracted to keep within a 1.5°C carbon budget" (Welsby *et al.* 2021). That is why there is so much emphasis on divesting from the fossil fuel industry and a prominent "KING/LINGO" (Keep it in the ground/ Leave it in the ground) civil society movement.

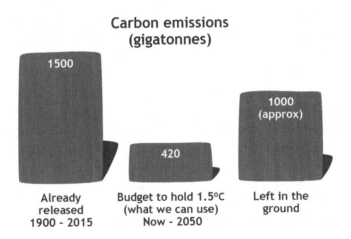

Carbon emissions (gigatonnes)

1500 — Already released 1900 - 2015

420 — Budget to hold 1.5°C (what we can use) Now - 2050

1000 (approx) — Left in the ground

Figure 2.1 Carbon: what we have used, what we can burn and what is left.

Back to the 420 gigatonne budget and the decade challenge. As of the early 2020s, we are still emitting carbon dioxide globally at a rate of around 50 billion tonnes per year. With a bit of simple maths, the challenge should be clear. At the current rate of use we have less than ten years before humanity expends its all-time budget and hits that 1.5°C of heating. Think about this budget like you might a household budget. You are spending recklessly with no thought for the future. Once your allowance is gone, that is it. There is no-one to lend you the money. You have to continue and deal with the consequences of your actions.

By the 2030s, then, we have to be well on our way to zero-carbon. By then, we will have a good sense of whether we have been able to dramatically reduce carbon emissions and avert run-away global heating. Clearly, the end of this decade is not some apocalyptic moment where global society either disintegrates or regenerates. The global climate system is not some machine with a dial that suddenly and uniformly shifts to "broken". Our natural world is complex comprising many different interconnected systems that can change in unexpected and very geographically uneven ways. What we can expect is that change will enhance existing vulnerabilities, finding ways to further damage places that are already fragile and precarious. Vulnerability is not equally shared. Many places across the world are already exposed to extreme weather events from climate breakdown, compounded by long-standing social inequalities, entrenched structural disadvantage from historic and current trade inequalities, as well as poor political management. What is important for our strategy is our ability to think in terms of a complex world system, put local events in a broader perspective of history, power and politics in order to understand and steer the future to a better outcome for all.

Welcome to the zero-carbon rollercoaster

So, if we are going to avoid the "spend and to hell with it" approach, we have to safely allocate our remaining budget. How do we make wise choices today to make the future safer? This concerns good carbon reduction planning – getting on a zero-carbon downward curve right now so we can transition away from fossil fuel dependency. Let's see how this works in practice for a particular city. We know that the global budget is 420 gigatonnes. If we take my own country, the UK, we can determine our share of that budget on population terms relatively easily by calculating the ratio of our population in relation to the world population, and then using that same ratio in relation to the global carbon budget. For the UK then we would have an all-time budget for burning carbon dioxide of around 3.5 gigatonnes (3,500 billion tonnes). The UK produces around 400 million tonnes of carbon dioxide per year, so it faces the same urgent pressure to put the hand brake on the rate of carbon emissions in about a decade.

We can focus on one place to see how this works in practice – in this case my home city of Leeds. There is a similar picture if we run the maths. Based on Leeds' population, we have an all-time budget of 40 million tonnes. With annual GHG emissions of around 4 million tonnes per year, Leeds has the same ten-year task. The planet, individual countries and specific cities all face the same challenge. They have to immediately get off the "business as usual" line of high carbon use and get on the zero carbon descent curve. Every year counts. To get to the half-way point in ten years, we have to see almost double-digit percentage reductions in carbon dioxide emissions produced by humans every year, not in the distant future but from this year onwards. Every year of inaction means higher cuts in the future. This would not feel like a gradual descent. It will feel like a rollercoaster.

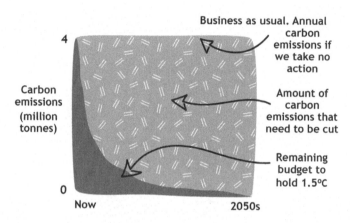

Figure 2.2 The zero-carbon rollercoaster

There are extra complexities that make this even more like a rollercoaster. At the moment most zero-carbon descent curves measure what is called territorial (or production) emissions – the carbon produced from activities in a given territory. This typically includes emissions that are easily identifiable from vehicles, home heating, activity in the local economy, as well as emissions from energy that is imported into that place. These are what we call Scope 1 and 2 emissions (Ghaemia & Smith 2020).

But there are many other activities in our lives that are responsible for high carbon emissions, and we are only just starting to think about how to include, and account for, emissions from flying, consumer goods and services that are imported into a place. In many high-income cities these can be a significant part of the story. Think about high street and shopping centres crammed with goods made somewhere else in the world, as well as the carbon-dependent travel that people in a place are undertaking elsewhere in the world. These activities mean that the overall carbon emissions burden of a place is typically much higher than those we can directly count within a place. This kind of budget can be called a city's "consumption footprint" (also called Scope 3 emissions), highlighting the total carbon emissions draw of a place to sustain life there. They are a better way of understanding city-based carbon emissions as cities are permeable entities, importing goods and services produced outside city boundaries (Barret *et al.* 2013).

In higher-income places, Scope 3 carbon emissions can be more than double compared to Scope 1 and 2. But remember, the carbon budget is still the same. Going back to the example of Leeds, this much higher level of use would "stretch" our budget and make the carbon descent curve really steep. It would make the ability of a city to quickly, safely and coherently reduce emissions much tougher, especially since it would need to tackle really tricky areas of flying and consumer spending, particularly amongst higher-income residents.

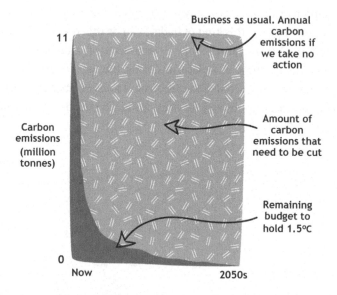

Figure 2.3 The steeper zero-carbon rollercoaster (based on Scope 3 emissions)

The final sticking point is the idea of fair shares (Friends of the Earth 2021). Up to now, we have been calculating these city budgets on a simple population ratio. The more people, the bigger the budget, and the opposite. But what about historic use, or indeed under- and overuse? A country like the UK has emitted far more carbon and enjoyed the benefits of this for many years, compared to many low-income countries in the global south. What if we adjusted those budgets for historic use? We have established that the all-time global carbon budget is estimated at 420 gigatonnes, and that means the UK gets allocated 3.5 gigatonnes by population alone. But if we look at that historic use, we can estimate that the all-time amount of carbon that humanity can produce to keep to that safe level is about 1,920 gigatonnes (the 1,500 gigatonnes already emitted plus the remaining 420 gigatonne budget). Based on current population the UK's proportion of this "all time carbon emission limit" would be 17 gigatonnes. It is estimated that the UK has already emitted around 80 gigatonnes of carbon emissions (Evans 2021) – several times more than expected based on population size. We are in fact, in deficit by over 50 gigatonnes. If we took an approach that adjusted for historic use the task is not to rapidly reduce the rate of our emissions, but to put carbon emissions in *reverse* as quickly as possible – a mass carbon removal exercise.

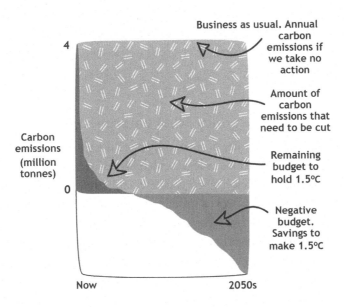

Figure 2.4 The negative zero-carbon rollercoaster, adding historic over-use

The same holds true for Leeds. By population size you would expect that Leeds would emit 204 million tonnes against that global all-time carbon budget. But an estimate of Leeds' historic carbon emissions since 1900 is 484 million tonnes. Leeds is actually in deficit by 280 million tonnes – using almost two and a half times the carbon budget given the population size of the city. This is the scale of the challenge for an old industrial city in a high-income country. The tragic part of the story is that the immense wealth that this created did not even lay the foundations for a city with a zero-carbon and equal future.

In reality, it is not likely nor sensible that such dramatic cuts could be put into practice. But it does highlight the kinds of geographical uneven-ness we are dealing with. Acknowledging historic uneven use – what we can call global carbon debt – and the way this shapes the present, will be essential if we are to respond to this zero-carbon city challenge in a socially just way. This requires taking a broader global and shared sense of responsibility, exploring difficult questions around compensation and reparations related to historic over-use as well as damage caused now and in the future.

Emergency, what emergency?

Let's now return to the idea of emergency that I use to frame action in this book. Dictionary definitions suggest emergencies are unexpected or dan-gerous situations, which happen suddenly and require rapid responses. The idea of emergency evokes strong emotions: fear, panic and concern for safety. It is a powerful term and should not be used lightly. There is nothing new in the idea of emergencies in cities. We are familiar with situ-ations that require urgent attention – a cry for help, a passing ambulance, a fire evacuation, as well as flooding, storms, heatwaves, landslides, fam-ines, civil unrest or military interventions. Dealing with emergencies has a more specific use in disaster management typically seen in lower-income countries, which lack the resources to deal with their impact.

Most places will have some kind of emergency disaster plan. In the UK for example, the Civil Contingencies Act (2004) provides a framework for disaster response, defining an emergency as "an event or situation which threatens serious damage to human welfare or the environment". Responses to emergencies come in many guises – evacuations, temporary shelters, emergency provisioning, curfews, rationing or blockades. But often there is also a flourishing of creative responses and community self-help in the face of emergency situations. We saw these kinds of measures during the Covid-19 lockdowns, and these offer glimpses of how we can become emergency first responders and develop our capacities for action.

As we saw earlier, the specific idea of a climate emergency gained momentum in the wake of the 6th Assessment Report of the IPCC in late 2018, which highlighted the urgent need for action to meet net zero carbon targets. Rather than something substantially new, this is the next step in longer standing calls for emergency action. Going back to the 1970s, landmark reports such as the Club of Rome's *Limits to Growth* (Meadows *et al.* 1972) and the creation of the United Nations Development Programme, policy makers, scientists, civic activists and enlightened politicians have been raising the alarm on the likely drastic impacts of climate change. In the ensuing years, we have had Earth Summits, scores of annual UN gatherings and declarations, new groupings and initiatives by global mayors, billionaires and celebrities. The phrase "last chance to save the planet" has been used more than once.

But since the 2018 IPCC report, debate and action on climate breakdown has subtly shifted. Rather than a technical issue for scientists, a cause for eco-activists or a preoccupation for policy makers, now there is a broader public awareness. As records are broken for the hottest year, the biggest superstorms, the largest wildfires, the longest drought or the mounting list of species extinctions, we now have evidence that the climate emergency is a lived reality for many, especially those least responsible (Wallace-Wells 2020). While there are still many voices who are delaying and denying, more people are now recognizing the emergency nature of the issue – that the climate is breaking and this presents a clear and immediate threat to humanity's survival as well as the mass extinction of species. Media and policy commentators, climate activists, politicians, even business leaders are converging in naming a need for emergency-type action. As if warnings could not get more severe, in 2022 the IPCC issued a further report highlighting the "increasingly severe, interconnected and often irreversible impacts of climate change on ecosystems, biodiversity, and human systems" (IPCC 2022) and stressed that the window of opportunity for transformative action would close by the end of the decade.

The last few years have also seen the rapid emergence of a language of climate and ecological emergency across public, policy and activist sectors. Greta Thunberg, Fridays for the Future and the Youth Strike movement took many people by surprise in 2018. It spread rapidly across the globe, and shook the self-confidence of many well-meaning adults. Back then, Greta, a 15-year-old Swedish school child, provided momentum for the climate emergency movement when she told the adult world that: "I don't want your hope . . . I want you to act as if our house is on fire" (Thunberg 2019a).

This shift was not limited to climate scientists or eco-activists. Municipal authorities throughout the world responded by formally declaring climate emergencies. In the UK alone, over half of its 308 local authorities have now declared a climate emergency. Globally, climate emergency declarations now cover nearly 900 jurisdictions and local governments across 160 million citizens (Gudde *et al.* 2021). In my own city, Leeds, I helped lobby local politicians to declare a climate emergency and commit to the cuts necessary to get us to carbon neutral by 2030. The task of declaring climate emergencies is only the first step. The hard work comes afterwards – developing a meaningful and costed planning process to implement emergency-type changes across whole urban areas.

The lack of meaningful city-level planning and action on the climate and ecological emergencies is now stark. The very idea of a climate emergency is a deeply challenging and disruptive framing for city policy and action. It takes urban policy far beyond established concerns with sustainability, energy saving or green infrastructure. Declaring emergencies is a dramatic turn for those trying to figure out the future direction of cities. The starting point is to match policy and action to the actual evidence, and to pick up pace. This might involve ditching or reworking much existing work and policy ideas. This is a real challenge. There is no clear definition of what we are up against, or how to respond, no forerunners or blueprints to learn from. We are indeed in contested and unknown territory.

It is understandable then, that the idea and prospect of a climate emergency can induce panic, fear and powerlessness. Does it really entail the complete transformation of life as we know it? Will it bring mass disruptions to food supply, energy provision, transport systems, changes in climate that leaves us at risk from constant extreme weather, crop failure, new diseases and sea-level rises? Will there be mass civil unrest, migration and more repression to ensure compliance with new ways of life? The reality is that no one really knows for sure. Different places will be affected differently. Those in the richer global north will use their existing resources to protect and shield their citizens from changes and continue with life as it is already known. Some places are already experiencing these effects. Wildfires in California and Europe, heatwaves in India and the USA, flooding in Pakistan, superstorms across Asia – all act as tasters for what are becoming more generalized conditions.

This brings us to a key point. Is the climate emergency really the big one that will engulf all others? The climate emergency is the new kid on the block and shifts attention from long-standing emergencies that have made up city life for decades, even centuries. For many indigenous groups, people of colour or those trapped in low-income lives, the climate emergency may feel like an abstract problem compared to those emergency situations that continue to shape their lifes – police brutality, racial discrimination, ableism, worklessness, domestic abuse, poverty, lack of access to clean water or shelter, gender oppression, inadequate healthcare. We have to be aware that to raise the alarm on climate can feel insulting to those who have been organizing against social disadvantage and oppression for many years but have been largely ignored. The climate emergency can feel like the preserve of a small, and more privileged, section of the global population who now want rapid change because it will affect them and their way of life.

The imbalance of action across different issues is a sobering reflection on long-standing social and spatial divisions in cities, which have become hardwired into life in a way that they become taken for granted and largely invisible. The daily drudgery of life for many seems so intractable that it can be ignored. More worryingly, there are city elites who would see divisions between people and neighbourhoods as part of the natural order of things, a price worth paying and a necessary part of how cities function. In this framing, inequality is not a structural feature of cities but a result of personal success or failure.

The broader issue is that we must not separate the triple emergencies – in terms of causes or effects. Indeed, when we say climate emergency, we need to move away from a narrow set of actions around carbon and energy savings. Our emergency framing is a way of thinking, talking and acting that can open up possibilities beyond our current way of life based on ceaseless economic growth, the legacies of colonial expansion, the plunder of resources, militarism, corporate monopolies and fossil fuel dependency built up over hundreds of years. The climate, social and ecological emergencies go hand in hand. They are part of the same story. That is why we need to talk about building a movement for climate justice as we respond to the climate emergency (Jafry 2019). For many people, if our responses do not create a better life where they can simply thrive and have a range of basic needs fulfilled, it is not an emergency response worth joining or supporting. Our emergency action needs to be broad, connected and relevant.

Building the 1.5°C city

Let's now look at our city task. The United Nations has provided our aim – to reach net zero carbon emissions as quickly as we can, and certainly be most of the way there in ten years from now. I call this the 1.5°C city. We have to create ways of city life that are compliant with, and do not exceed the Paris Agreement target of holding average global temperature rises to no more than 1.5°C above levels seen at the beginning of the industrial revolution. That is a huge task. All activities, everywhere, by everyone, for all time, have to be, at minimum, 1.5°C compliant. City-level action will have to be rapid and transformative across a range of areas including transport, housing, food and diet, waste, industry, retail, planning, land use and education, creating percentage reductions in annual carbon emissions in double digits and shifts in policy and activity patterns without historical precedent. While previous world events have displayed the kinds of carbon reductions we need – the Covid-19 lockdowns of 2020, or Russia after the collapse of the Soviet Union – these moments came at huge social and economic costs and certainly did not give rise to a more equal society. Given the scale of the task ahead, we have to move quickly but fairly. We know that we have to do it in a way that leaves no one behind.

The task has to start immediately. Delaying only makes our ability to reach net zero even more difficult. We need a clear idea what this means for cities; how early climate emergency declarations can be translated into meaningful civic action and public policy, with identifiable year-on-year targets and noticeable changes in daily practices. Exactly how much carbon, what changes, by when, by whom and how? These are the questions we need answers to, and where resources and insights are required.

This is complicated on a number of levels. First, the scale of the task requires system-wide redesign, rather than local or targeted interventions that tinker with bits of city life (Madden 2019). As I explore in our strategy chapter, we need to think of city life as a system of interconnected elements – how we consume, plan, produce, finance, distribute, advertise, own and govern. It is no longer sufficient to keep making small adjustments to bits of these systems. What is required now is a redesign of whole city systems. This task is unfamiliar to those involved in city governance, planning and policy making, but as I explore in subsequent chapters, it is a task that they urgently need to take on.

Second, the level of change required not only alters the scale and direction of particular city systems, it also changes how they function. To take the example of city mobility, there will not only be fewer journeys made by cars, shorter food miles or more local supply chains, but a new system-level logic to why we need to move around. As changes become deeper and more widespread in the 2030s, city systems will start to functionally change internally and in relation to each other. This is difficult to understand or imagine from the current model of globally-connected high-growth cities in which city systems function in a particular way. Once they are decoupled from a fast-growing external economy, and recoupled to a more localized, steady state and circular economy, a new purpose for our city systems come into view. Nature, climate, resilience, equality, dignity, well-being and quality of life become central concerns.

Third, cities are not level playing fields. Cities with higher levels of carbon emissions need to act quicker and more boldly. Those with historically lower emissions, especially in the global south, should be allowed to reduce their carbon use more gradually, especially through support and funding for zero-carbon technologies and infrastructures, and converge around a common safe point (Meyer 2004). Many places need to be allowed to make up for structural inequalities in global trade and politics that they have faced for hundreds of years. While we have to achieve an average citizen carbon emission footprint of around three tonnes of CO_2 equivalent per person by 2030, what is clear is that many city-dwellers across the world, mainly in the global south in parts of Sub-Saharan Africa, Latin America and Pacific-Asia, already live below this. In contrast, richer established cities as well as rapidly expanding ones have emission profiles which if left unchecked commit the world to well over the 1.5°C danger zone of global heating. So, while there is a shared global endpoint, the task will be very different in different places.

Fourth, rapid GHG reductions have to deliver a range of what C40 Cities called co-benefits (Floater *et al.* 2016). These are win-win situations, where one benefit creates a tangible benefit in another area. Think about a shift to walking which instantly improves city air quality and has a public health bonus. Changes that work towards the UN's Sustainable Development Goals are especially useful. Goal 11 focuses on: "making cities and human settlements inclusive, safe, resilient and sustainable" (UN 2022). This helps us place our rapid city carbon reduction activities in a commitment to social equality, and historically rebalancing who has benefited unequally from growth. In Chapter 5 (Moves) we will focus on these win-win situations through activities that positively impact the triple emergencies. Our city emergency action on carbon reduction and nature recovery also needs to address long-standing urban problems: poverty, segregation, planning blight, air quality, environmental degradation, uneven land ownership, precarious work, housing, racial and gender oppression and citizen disengagement.

Fifth, confusions can emerge over definitions and timescales and these have important effects. Many cities have declared climate emergencies but crucially three different framing are used: carbon neutral, net zero and absolute zero. The first two are similar. Carbon neutral means the amount of carbon emitted is equal to the amount absorbed from carbon sinks like forests, soils and oceans. Net zero is the same principle but refers to all greenhouse gases and not just carbon (University of Oxford 2020). The issue for city change-makers is that carbon neutral and net zero targets can open up short-cuts and misleading actions through emissions offsetting and trading. Carbon savings can be bought or traded from other locations or activities. Typical offsets might include investing in green energy or tree planting. Such approaches offer cities a get-out for their carbon-saving responsibilities allowing slower and less intense changes, and importantly for life to continue as before. Moreover, a large global market has emerged to trade offsets so it is often difficult to trace if the actual savings happened (Childs & de Zylva 2021). That is why activity that actually reduces carbon emissions within cities is the most accountable route and also offers a host of co-benefits for citizens.

Absolute zero is more challenging as it aims to eliminate all sources of GHGs, with the proviso that there will be some "residual emissions" that cannot be eliminated easily. Negative emission activities that actively sequester carbon address this residual element, and these come in various forms such as large-scale geoengineering options as well as more localized, nature-based solutions. In fact, the longer we leave it, the more we will need to rely on these negative emission activities along with actual carbon reduction measures.

Looking across city declarations all of these approaches are used with different timescales ranging between 2030 and 2050. The key issue is that there is no universally agreed approach to adopt. No-one knows for sure what is safe and achievable. We just need to act faster than we currently are doing, and lock-in beneficial social and ecological changes as we do this. As momentum, and a sense of unease, grows, we will see deadlines compressed and targets brought forward. Many cities have the ability, resources and leadership to create roadmaps that can get to zero carbon earlier than the 2050 target. But the important point is to avoid choosing targets and timescales to make the task more or less easy. What is important are the immediate changes we can embark on, this year and next, and that our efforts aim for broader system change at the city level.

Finally, the big message for our efforts to save the city is to measure all our activities against social justice. After all, how can we justify the creation of a zero-carbon city if we simply maintain the same social, gender or racial inequalities? Globally, the picture is stark. The average citizen has to reduce their emissions to almost three tonnes of CO_2 equivalent by 2030 – to put this into comparison the average for North America is over 15 tonnes, Sub-Saharan Africa is less than one tonne, and Latin America is around two and half tonnes (World Bank & Climate Watch 2020). The inequality is clear and it is growing. The activities of a small minority of the world's population is bringing our biosphere closer to collapse, while a majority struggle to get the basics for life. This is not just a story of global division. There is extreme poverty in many high-income cities, as well as wealth and luxury in lower income ones.

How will you spend the decade of transformation?

Going back to our decade-long challenge, what are we supposed to do? If we have roughly until the early 2030s to make a difference on how the rest of the century, and the fate of humanity, turns out, what can we each spend our time doing from now to make the most difference? We can begin by reviewing the resources, networks and influence at our disposal. In this book, we will look at the different players involved in the change game. Which players do you reflect, or which would you like to be, or associate with? How can you mobilize action in your family, workplace, college, place of worship, or even the shopping mall? How can you contribute to saving the city from the place you stand? In the pages that follow I offer insights into the strategies, players and moves we can make.

In my case, I am an academic, writer and researcher in one small, and relatively privileged, corner of the world. The question academics can ask is what teaching and research is commensurate with the scale of the challenge? In ten years, a tenured social science professor at an international university might publish a dozen papers in leading academic journals. They might get a major funded grant, give some conference presentations, review other academic work, examine some students and in the process stack up a good amount of international travel and associated carbon footprint. Some satisfaction, prestige, intellectual gratification and career progression would result from these.

But this is no longer enough. Perhaps it never was. The publications might be read by a few hundred people, couched in technical and disciplinary language, sat behind corporate paywalls accessed by a privileged few, in global terms, at universities and research centres. But the academic publishing industry, as well as much of the agenda of government sponsored research councils are increasingly aligned with a globalized and corporate-led growth model (Castree & Sparke 2000; Pusey 2017). Priorities commensurate with the huge challenges of societal transformation ahead remain fringe. The triple emergencies need to take centre stage and become the serious business of a reimagined university. I decided to do what I can – "to dig where I stand" – and write a book that gets to the heart of what I see as the decade's challenge ahead and how to immediately get into emergency mode. As an urban geographer, my preoccupation is how to fundamentally change city life for the better. By the 2030s I want us all to have access to a prosperous, safer, shared urban future on this finite planet.

Learning from the Covid-19 city laboratory

Reflecting on the last few years, I explored the coronavirus-induced lock-downs as real-time laboratories full of living examples of what a more sustainable future might look like (Chatterton 2020). While this period was incredibly harsh and indeed deadly for many, it was an opportunity to study and consider which activities could be (re)used to build sustainable, and safer, cities. What can we learn from these crisis-led lockdown innovations as we attempt to save the city? Many previously unthinkable things became possible. Rapid changes were unleashed to recast the economy, health, transport and food. Fragments of progressive urban policy emerged: eviction cancellations, nationalized services, free public transport and healthcare, support for rough sleepers, sick pay and wage guarantees, localized services and a better work–life balance. There was a flourishing of community-based mutual aid networks as people volunteered to help the most vulnerable with daily tasks. If people get organized these gains can be protected. Yesterday's radical ideas can become tomorrow's pragmatic policy choices.

As we emerge from the worst of the Covid-19 pandemic, a key issue lingers: what is a city for? Is it to pursue economic growth, attract inward investment and compete against global rivals? Or is it to maximize quality of life for all, build local resilience and sustainability? These are not always mutually exclusive, but it is a question of regaining balance. Beyond politics and ideology, most people simply want to be safe and healthy, especially in the face of future threats, be they climate, weather or virus related.

I contend that the aim for cities now is to respond meaningfully to the converging emergencies of our age and urgently set up mechanisms and processes to begin substantial debate and action on the decade ahead. There are often limits on the powers and resource of cities in very centralized countries, and these limits need to be addressed urgently. City plans need to be supported by national climate emergency plans, with a commensurate transfer of resources and powers. There needs to be alignment between city and country through favourable national primary legislation to fundamentally shift the allocation of subsidies, infrastructure planning and taxation.

The change journey ahead will require understanding, listening to and connecting with the real and everyday concerns that people have in the context of rapid change. City leaders and citizens need to come together and jointly own the actions and benefits associated with changes. While this is an emergency, it has the potential to have deeply positive outcomes, addressing many of the stubborn problems that have plagued cities for generations.

No particular urban future is inevitable. The future story, and reality, of our towns and cities is up for grabs. The positives that were glimpsed during the Covid-19 pandemic could feasibly be locked in and scaled up to create a fairer, greener, safer urban future. We saw that we can all live well, and even flourish, in cities even if we have and do a little bit less of the things we have become used to. Revaluing what is important – community, friendship, health, family life – allows us to see how much we already have that can improve our well-being. From here on, every year counts. We need to think big, start small, but act now.

* * *

The book now explores how some of this might happen. I start by looking at our strategic approach to orient us in these complex times. Our guide for emergency action starts with the recognition that we have to learn and unlearn, act and resist, and make and build.

3

Strategy: our approach to change

How does change happen?

So, now we have a better sense of the challenges, and what we are up against, how should we go about saving the city? Or put it another way: how do we think change happens? This is an important question. We live in a complex, vast and messy world, where billions of people are simply trying to get by, improve their lot, and make sense of it all. All of us, to different degrees, have a sense of how we think the world works and how change happens. This is built up from our own experience, from what we have read, and from what others tell us. Most people, then, are in the change game. There is, of course, the influence of lobbying, manipulation, disinformation, evasion, corruption and deception. This is why engaging with how change works is so important.

The reason I start with strategy, is that if we are going to influence and change our cities positively, there needs to be some time dedicated to think about how and why change happens, and importantly how we link our aims and outcomes. What interventions do we need to undertake to get from where we are to where we want to be? Are we basing our work on causal links between evidence, action and outcomes? This kind of method is called a theory of change (UN Development Group 2017). In reality, there is no one-size-fits-all approach. There are multiple and competing theories of change. It takes a while to build up a clear sense of how and why change happens. Groups can be more or less open on the topic of whether they have a theory and how they see change happening. But anyone who wants to influence the future needs to adopt some kind of approach to change. There are no guaranteed formulas. What works in one time and place might not work so well in others. Some of this depends on luck and circumstance, as well as hard work and perseverance. Success also might only become clearer with the passing of time, after a group or initiative has ceased. Much of our change work, then, is about contributing to the flow of ideas and energy passed down over the years rather than immediate wins in the here and now.

Many different people, groups, organizations, institutions and political parties have developed a theory of change over the years. These differences depend on a range of important factors, especially in terms of your relationship to the three significant building blocks of our world: the state, the economy and civil society. Those in government and statutory agencies may see the role of the state, both local, regional and national, as the route to change, using taxes and regulations to provide welfare, keep their citizens safe and creating conditions for prosperity, investment and a good life. Ultimately, the state has vast powers through its regulatory, legal, policing and judicial systems that, as we saw during Covid-19 lockdowns for example, can rapidly set the terms for behaviour and daily life. As a last resort they can also make change through military means, often with deadly and tragic consequences.

Those in business, both large and small, are more likely to regard exchanges in the market determined by prices as the key driver of change and the route to prosperity. Clearly, there is a huge variety here. The diverse economy approach of Gibson-Graham (2008) highlights that the realm of the market encompasses everything from barter and family exchanges, co-operative and employee-owned enterprises to mega transnational corporations. Lastly, there is civil society – the family, trade unions, faith, voluntary and community sectors as well as social movements. This building block is often overlooked, but in many ways is the source of innovations that drive everyday well-being and underpins our emergency responses. The relationship between state, economy and civil society is critical (Wright 2011). Different combinations emerge in different places at different times. For example, a strong state can enable civil society and create a more socially responsive economy; a weaker state can leave space for a strong corporate private sector to flourish and social conditions that are heavily shaped by market priorities; while a strong civil society can push for, and protect, social and ecological gains.

Learn, act, build

I have developed a particular approach to city change based on my own experiences, inspirations and actions over the years. I present it through an approach I call: learn–act–build (LAB for short). The three elements work together as a connected set of ideas to create depth, momentum and focus. There is a strategic logic to what I propose that is worth following in a particular order.

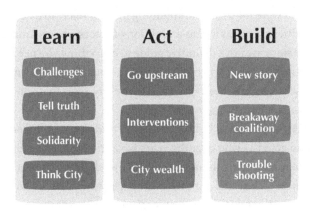

Figure 3.1 The learn-act-build approach

The first step focuses on learning and, importantly, unlearning. To effectively create change, we need to learn new ideas and ways of thinking, as well as unlearning established ways of seeing and acting that hold us back. These stem from the unconscious biases in our everyday lives as well

45

as broader structural factors that shape how we see the world. This learning and unlearning is not a one-off process. The most effective approaches to change are constantly learning and adapting. There are several key aspects to this first strategic step of learning and unlearning including recognizing the challenges ahead, telling the truth about them, developing solidarity with those most affected, and importantly, starting to see change from the perspective of a city.

The second part of the city change strategy relates to acting and resisting. Applying what we are learning and unlearning, how do we empower ourselves and others, take action and push back against what is harming us, especially in our emergency context? In this strategic step I explore the pinch points for our actions, how we need to link our immediate activities to broader issues upstream, and how we need to act against the urban growth machine to create city wealth. But we cannot just resist; we also have to create. So, our final part focuses on building and making, creating new ways of being and living in the present that can support a viable and prosperous future for us all. This building process refers to a new story for what our cities are for, as well as a coalition of actors who will make it happen. These steps are not a definitive guide to everything you need to know. Use them according to where you are at in your change journey. Think about these strategic issues as the rules of the city game that you can try in your role as emergency first responder.

The LAB approach is built from a rich tradition of thinkers, practitioners and activists who over the years have tried to make change. First, Joanna Macy and Chris Johnstone's (2012) work on the "Great Turning" stressed three connected aspects of change work: actions to slow the damage to Earth and its beings; an analysis of structural causes and the creation of structural alternatives; and a shift in consciousness and the deeply ingrained values that we hold. This is a really inspiring tripartite way of understanding the need to work at multiple levels and intents. Similarly, the Ayni Institute (2021) has undertaken work they call "Movement Ecology" which highlights different but connected ideas that shape our work including: personal transformation work we do as individuals;

creating alternative institutions that make the world we want to see; and institutional change based on resisting and challenging dominant power. Second, Bill Moyer, US activist and writer, developed the "Movement Action Plan" (MAP) (2001), initially in the 1970s, as a strategic model for social movements that uses eight distinct stages, and shows how to effectively match tactics and strategies to each stage. Third, work by the Sheila McKechnie Foundation on social power (2018), highlights strategies for maximizing the capacity of civil society to deliver transformative change through a range of more collective (strikes, assemblies, protests) and individual strategies (storytelling, enterprise, advocacy). Fourth, the US-based Center for Story-based Strategy (2020) focuses on movement-building using the power of narrative for social change, an essential part of building a new city story.

Fifth, my perspective owes a great debt to the work of the late Eric Olin Wright (2010) who developed ground-breaking emancipatory social science in the United States, part of which highlighted three interconnected strategies for social change: ruptural (breaking free and pushing back against the present), interstitial (developing alternative projects beyond the status quo) and symbiotic (working inside and with existing institutions to develop change). What we can take from Wright is how these combine in our change work. We need to work in, against and beyond the present, creatively reusing but also innovating from what we find. Finally, Kate Raworth's "doughnut economics" (2017) approach, downscaled to the city level through the Doughnut Economics Action Lab (2021), encourages us to explore how we meet local aspirations while also meeting our global responsibilities. Importantly, it asks how can a city's inhabitants move towards the safe operating space of the doughnut where we do not exceed ecological limits or drop below a minimum social foundation. This is a really powerful metaphor to guide our work. Many aspects of city life are currently outside this safe space of the doughnut, whether it is how we treat local nature, the social safety net available to people, our impact on wider planetary health or our impact on people across the world through forced and child labour or unsafe working practices.

I have experienced and used aspects of these approaches in a range of activities, past and present, including supporting camps for Climate Action resisting the fossil fuel industry, living and working with the inspirational Zapatista autonomous communities in Chiapas on Mexico's southern border (Chatterton 2017), running popular education workshops in schools, colleges and universities with the Trapese Collective (2007), helping set up housing co-operatives, creating independent social centres, building a low impact neighbourhood (Chatterton 2015), running a Masters course on "Activism and Social Change", leading a legal defence with 28 campaigners protesting against coal-burning power stations (Monbiot 2009), and supporting climate emergency action with a civic coalition called Our Future Leeds. What I take from these experiences is that our approach to change requires a range of tactics and actions at the same time. We can distill this as the need to create and resist, to learn what we are up against, joyfully construct the world we want while pushing back against what harms.

Below, I take a more detailed look at our strategic steps, grouped under the three parts of the LAB model.

Learn and unlearn

The first group of strategic steps focuses on learning and unlearning. Throughout our lives, we all learn about how cities work. We silently absorb the rules of the game. This is useful. On one level it allows us to function and stay safe. We figure out where food comes from, how you access housing, where is good for leisure and sport, which areas to avoid. Much of this is derived from experience, as well as family, peer group and social media. We come to know the city in a particular way. These understandings are always partial, and can be based around prejudice and assumption as much as reality.

Our task is also about unlearning. There are many aspects we have learned that hinder progress, stop us seeing solutions, and importantly block meaningful relations with others. One of the most significant aspects is how much of our learning has come from dominant institutions and social groups, which over the years have reinforced that there are particular ways of thinking and acting, and that these are somehow natural and offer better outcomes. We have built up a powerful set of unconscious biases which filter the way we see the world, especially in terms of class, race, gender and ableism. Our task is to unlearn those patterns that continue to block change, harm us and others, and learn new and often uncomfortable patterns of thinking and acting (McArdle 2021) that can take us beyond the business-as-usual of city life. We have to be realistic about the multiple barriers to a different approach to learning, from peer groups, our own inbuilt protective systems that filter the world in the way we want to see it, as well as poverty and inequality that forces many people into a task of survival. Let's have a deeper look at this learning and unlearning.

Identify the challenges

If you are responding to an issue, a good place to start is knowing the scale of the challenge you are up against. This is usually how we run our lives. If we hurt our finger, we don't call for an ambulance. But if our house catches fire we call the fire service. If we see someone driving dangerously we might get annoyed, but if a car mounts the pavement hopefully we try and jump out of the way. Perceiving danger and knowing how to respond is part of being able to live safely.

Knowing and recognizing the scale of the challenge is a crucial first step in saving the city. But the task is complex when the challenge is less visible. We cannot see carbon accumulating in the atmosphere, and most of us will never engage fully with the science of climate change. We need to be able to access a basic level of education that equips us with a critical ability to learn, interrogate and understand from others that there is something called the greenhouse effect that is heating our planet to dangerous levels. This task is made more difficult by a well-organized misinformation industry that exists to purposefully question and dispute the scientific consensus on rapid human-made climate breakdown (Hoggan & Littlemore 2009). The fossil fuel sector is largely behind this, which is understandable given they have the most to lose from their vast amount of sunk capital in the extraction of oil, gas and coal.

However, the climate emergency is becoming more visible at an everyday level through disruptive extreme weather events such as flooding, heat waves, droughts and fires. The longer term knock-on effects of these are a whole basket of consequences around disruptions to food and energy supply, loss of biodiversity, the spread of diseases, and the displacement of jobs, people and other species (IPCC 2022). The social emergency of inequality, poverty, oppression and precarious lives is also a constant presence. New vulnerabilities layer on top of existing social inequalities creating people and places deeply unprepared for the future (Chancel 2020). This coming together of our nature, social and climate emergencies are classic wicked problems (Incropera 2015); causes and consequences become deeply difficult to unravel, and they start to feel so ingrained that they are accepted as inevitable.

Identifying challenges, then is not easy or straightforward. It requires patience, time, access to resources and learning opportunities. It requires a commitment from the state at all levels to offer affordable and lifelong learning opportunities so we can all develop an ability to think critically and independently. Often it is about unlearning what we have already learned – questioning assumptions as much as learning new facts and figures. Compounding the problem, many institutions and educational establishments are not up to the task. This recognition was one of the driving forces behind the emergence of the youth strike movement. Thousands of young people started to question why they were not being taught about the scale and severity of the climate emergency at school and college. As I explored in the last chapter, we have to understand what climate breakdown is, but more importantly what is causing it. The climate problem is often presented as something separate to us, as a geophysical malfunction of our Earth system – the climate system has gone wrong somehow and needs fixing. Rarely is it presented as a result of broader economic and historical trends with roots in military conquest, colonialism and a capitalist economy based on extraction, enclosure and commodification. To locate it in this way helps us to sharpen our focus on the way of life of a minority of the world's population and how they are driving unsustainable levels of energy use, consumerism, pollution and carbon emissions. The inequalities of responsibility are stark. A recent report by Oxfam highlighted that the richest 10 per cent of the world's population were responsible for just over half of all carbon emissions in the past 25 years, while the poorest 50 per cent of the world's population were responsible for just 7 per cent (Gore 2020).

Focusing on inequality shifts our perception of what we are up against and what we need to do. The challenge is to confront and disrupt the distribution of power and resources. If climate breakdown remains externalized as something in the atmosphere we cannot bring the broader patterns of deeply unequal city life, and what creates and maintains them, into the explanation. But even when we do start to shift our learning (and unlearning) there is a further challenge. We have to match our solutions

51

to the scale and speed of the challenges. Health researchers talk about the "evidence-action" or the "know-do" gap (WHO 2005). This work highlights that there is often a lag between what we know and what we do. It simply takes time to catch up with the latest evidence and put it into action. In an emergency context this can be an issue. We need quicker feedback loops between what is happening and how we respond. A key part of our response is remaking our governance and educational institutions in a way that can close this gap.

There are other reasons why this gap persists. Political inertia and not wanting to accept the scale of the challenge is certainly one. But also many of us operate in contexts where it is difficult to bring forward solutions that are needed. This works on many levels. For a solution to be rapidly and broadly adopted it has to navigate the complexities of how city systems work. We have to ask is it technically possible, financially viable and socially acceptable? It is very difficult to meet all these criteria, especially given subsidies, taxes and regulations that still support carbon emitting activities. We still largely operate in a cost–benefit model, where only solutions that can demonstrate cost-effectiveness succeed. As we confront the immense existential threat of the end of safe conditions for the human species under circumstances of its own making, we have to ask what price we place on survival. Will we really be the first species to not take emergency action because it wasn't cost-effective?

Tell the truth

What we do with our insights into the challenges we face is the next big step. The challenge of our triple nature, climate and social emergencies has to be told, and retold. It has to become a popular narrative that creates a groundswell for broad and popular action. Extinction Rebellion (2022) frame this challenge as the need to tell the truth and put it like this: "We are facing an unprecedented global emergency. Life on Earth is in crisis: scientists agree we have entered a period of abrupt climate breakdown, and we are in the midst of a mass extinction of our own making".

While this is an emotive framing, the intention is the important part. Many people are unsure of the nature and scale of the challenge, some are simply in denial, others might be confused or seek to confuse. The task in this context is to gather and understand the facts and confidently repeat them – tell the truth to those who will listen, and especially those who hold power and control resources (Alinsky 1971). The idea of speaking truth to power has a long history in the peace, Quaker and civil rights movements, mainly as an alternative route to direct violence. It can be an empowering act, and a moment where people with less power hold those with more power to account. This task is made more complex in an era of fake news (O'Connor & Weatherall 2019) where multiple and competing truths circulate, and who provides information becomes as important as what is said. Moreover, for those who might not have the ability to defend themselves or live in contexts where free speech is not protected, telling the truth can be dangerous, even deadly.

Our learning and unlearning work can draw on longer traditions of popular and community education to sharpen and spread understanding of the challenges we face. Drawing on work of activists and educators such as Myles Horton and Paulo Freire (1990), popular education is a method through which those with less power can understand what they are up against, and how to act and organize to gain more power. Popular and community education can play a role in creating conversations in workplaces, government departments, schools, places of worship and canteens across our cities. Through workshops, teach-ins and training manuals they can offer clarity on the current state of play and what we can do across the converging emergencies. Exciting innovations are also spreading, including citizen juries and assemblies which bring randomly selected people together to explore issues and come up with recommendations to respond to them (Bryant & Stone 2020). Telling the truth is about breaking ranks, questioning long-held assumptions, simply saying "I am not going to do this any more". This could involve big or small acts – challenging planning decisions, changing the way contracts are issued, who we vote or speak out for, who we see as allies, what we choose to read, or how we speak to our neighbours. Almost every act matters.

Part of telling the truth is about interrogating whether our proposals and actions meet the challenges. Do they mean rapid decarbonization in the next ten years, that will also ensure more social justice and safeguard our natural world? In Chapter 5, I explore a range of emergency moves and frame them in a way that is up to this triple emergency task. Telling the truth is not an easy or everyday skill. Communicating effectively requires effort and training. Pioneering communicator Marshall Rosenberg developed an approach to non-violent communication to do just this (Rosenberg 2015). It starts with understanding where we and others are coming from, listening with empathy and using that to build genuine dialogue and understanding. Ultimately telling the truth is a two way process. It is about finding a way to frame the challenges in a way that makes sense to a broad group of people so they also feel empowered to act.

Climate justice: solidarity and social justice

The third step asks a simple question: as we respond to the climate emergency, are we also responding to the longer, persistent social emergency (Hulme 2019)? Are we creating a fairer, more just as well as a climate safe and greener world? Tackling the climate and ecological emergencies has to be an opportunity to also tackle poverty, oppression and violence as well as a range of inequalities related to gender, race, physical ability, sexuality or neuro profile (Hopkins 2018). Many people are simply not going to support changes if they do not improve their lot. We forget this at our peril. Without real attention, what might be considered pro-climate changes can actually harm daily life for some. Consider electric cars that increase the extraction of heavy metals and the displacement of communities whilst also undermining walking and cycling, or new digital platforms for home heating or local transport that exclude those without access to up-to-date mobile phones. Without serious intervention aimed at social equality, improvements are often left to the market, accessible only to those able to pay. This strengthens the position of large corporate entities and creates a tiered society between those who can and cannot access the benefits of a green transition.

In many ways, then, saving the city is a social change project. Making social changes is the route to creating positive outcomes in terms of the climate breakdown. We need to ask who the city is for? There is a crucial issue of ethics here based on solidarity and internationalism – a desire to care for others and make connections with those trying to build a better world (Featherstone 2012). These are not easy ethics to embrace. Many cues in our divided and heavily consumer-focused world push us towards self-interest and self-gratification, while the sheer precarity of modern life traps many people in the realm of immediate survival. But in the face of these emergencies, we stand and fall together, locally and globally (Routledge & Cumbers 2009). Building transnational networks of solidarity will be key, reaching out to those across the world to put the ideas behind this book into practice. The watchword is interdependence; not just between humans, but between us and the natural world we depend upon. The city you live or work in is not an island. It is a deeply interconnected place that has impacts on people across the planet. Our ability to reach out, understand, and act for other people will be one of the defining factors of how we handle the coming emergencies.

Giving voice to groups traditionally excluded from power in this process is essential to ensuring social justice and compassion are built into changing the city. It is a way of understanding people's lived experience of oppression and harm, and also why many people are stuck in ways of living that are harmful to themselves, others and the planet. Focusing on solidarity also means understanding and reversing the harm that we might do to others, nearby or elsewhere, unintentionally through our everyday consumer habits. This is a broader agenda for the right to the city, and making this real in the everyday (Harvey 2013). Complexities arise here – how, for example, the right to consume or drive can harm the right of others to lead a safe and dignified life. The right to the city is a broader ethics where we acknowledge and act on our interdependence with others, both human and non-human.

The idea of climate justice is central here (Robinson 2018). This term essentially asks us to see our actions on climate change as a way to work towards greater social justice. At its heart is the unequal story of colonialism and climate change (Barber 2021). Those least responsible for climate breakdown are today dealing with its worst effects, and have less access to resources to protect themselves from present and future harms. People and places continue to be impacted in very different ways. Think of coastal cities in Asia susceptible to flooding, low-income groups unable to protect themselves from extreme weather, subsistence farmers vulnerable to crop failure. Many places structurally disadvantaged in the global economy are not able to access the resources needed to build resilience in the same way as richer nations.

Left unchecked, then, climate breakdown will increase harm, injustice and vulnerability for those least responsible for it. Vulnerability is layered on existing vulnerability, creating a downward spiral of mounting problems. We need to make sure actions to address the climate emergency build in social justice and directly unravel these existing inequalities that make up our world. The biggest threat at a city level is further segregation and division – where those with more resources retreat into walled ghettos and enclaves, relying on a mix of technological surveillance, private security and gated communities to defend their lifestyles and resources (Graham 2010).

The climate justice movement originates from places where climate breakdown has been impacting people's lives for decades. Pacific islanders, Indigenous peoples and subsistence farmers are the frontline communities in this struggle for climate justice (Dhillon 2022). Whether it is dealing with sea-level rise, strip mining, fossil fuel extraction, or the expansion of highways and pipelines, their daily lives are subject to displacement, violence and pollution. Such tragic events need to cease simply to reduce human suffering. But there is a further reason. While Indigenous peoples make up about 5 per cent of the world's population, the land they inhabit contains 80 per cent of the world's remaining biodiversity (Recio & Hestad 2022). Indigenous peoples are key to the survival of humanity. They are the stewards of our global ecological system upon which we all depend.

A climate justice approach, then, is part of a broader project of decoloniality: to think about, and act on, how deeply damaging and violent colonial legacies still work in the present (Mignolo & Walsh 2018). Climate breakdown has very uneven causes across time and space. Not everyone, everywhere contributed to this situation equally. Some nations and social groups in the so-called developed world for example, have accelerated climate breakdown through historically higher levels of fossil fuel use, while others across the global south, have hardly contributed at all. We have to recognize that climate breakdown emerged over centuries as European nations brutally colonized large parts of the world, enslaved people, enclosed lands, extracted resources, established deeply uneven patterns of trade and development and developed an industrial system dependent on fossil fuels, based on continuous economic growth and the production of consumer durables which ultimately spread to a planetary level (Gunder Frank 1966; Nakate 2021).

There is an important racialized element to the climate emergency (Williams 2021). Like many aspects of our cities, people of colour are disproportionately affected by climate breakdown. This is of course not due to any pre-existing characteristics of particular people or places. It is a reflection of the structural inequalities and uneven spread of resources and vulnerabilities that have shaped our world for centuries. The Black Lives Matter movement that grew in international prominence in the wake of the murder of George Floyd, has provided an essential focus to understand this. Importantly, the idea of white supremacy has spotlighted how racism works on a largely invisible but structural level (Saad 2022). A white, and also largely ableist and male, culture creates a dominant centre that becomes a reference point for how our world should or ought to be (Wa Thiong'o 1992). Anyone outside this dominant culture has extra work to do just to keep up. Importantly, in times of crisis tragically it is those from the dominant culture who are able to mobilize resources and greater levels of protection. Hence, the cycle of structural violence reproduces itself. Saving the city means breaking this cycle, shifting the dominant centre, and providing the conditions for flourishing and safety for everyone.

Understanding how our cities are shaped by the dominant culture is a huge and destabilizing challenge to those who benefit from it. Just by being part of the dominant culture is to maintain some level of injustice. This is no time for guilt or blame, but for relearning how our world works so we can show allyship, solidarity and internationalism to tackle the converging emergencies. As I explore later there are many ways we can put these approaches into practice in terms of how we design, manage and own our cities.

Think like a city

Now we arrive at the last step of our learning and unlearning: think like a city. This might sound like an odd step. How do you think like a city? My point is that we need to take a helicopter view of how cities work: to see them as deeply interconnected places where daily life is messy and a product of interactions between different groups and interests. This perspective is essential to avoid a naïve localism (Clarke & Cochrane 2013), an approach which regards places as hermetically sealed from the outside world and which overlooks how politics, power and money shapes them. One of the most limiting aspects of this localist view of cities is that the problems they face are seen as products of that place, and entirely solvable from within. However, the reality is that places are made up from complex flows and networks of ideas, people and resources that move in and through places (Massey 2004). This makes place-based action more complex than at first sight.

Indeed, simply pouring resources or policies into places does not work in a straightforward way. Issues like disinvestment, crime, poverty, climate breakdown and biodiversity loss are complex and cannot simply be turned on or off in places. Priorities in corporate boardrooms, events in national parliaments, carbon accumulating in the atmosphere, animal species slowly declining or migrating are all events that unfold around us and deeply shape our lives but are out of sight and influence. We also have to avoid what Mark Purcell (2006) called the local trap – assuming that cities simply represent a better kind of politics than other scales such as the region or the nation. The reality is that we should not idealize the city. It is subject to all the same faults, difficulties and errors that we find at other scales. Rather, our focus should be organizing and taking action to support the kind of city we hope for (Russell 2019) and desperately need to tackle the converging emergencies. To save the city, then, we have to think more broadly about causes and origins across multiple scales and temporalities.

For example, while cities are a collection of material things – buildings, roads, pipes, plants, trees, soil, rock and people – they are much more than this. They are also made real through relationships between people that can make things happen to a greater or lesser extent. Often the success of these relations is determined by how much they are networked and connected beyond places. The sociologist Mark Granovetter (1973) called this "the strength of weak ties". Cities can build up power and resources to act, but these also depend on having extensive influence, contacts and relations beyond city limits. To think like a city we have to see our urban places as deeply interconnected entities that rely on flows of people, animals, resources and natural features such as water, air and soil to function and ultimately to flourish. What is apparent is that places are equipped to a greater or lesser extent with what they need to fulfil the needs and desires of people in that place.

The task, then, is to harness this diversity to create a different kind of change process. Cities are full of innovators and change makers setting up new projects and establishing new ways of thinking. This might be techniques for food growing, making renewable energy or ways of getting around. The power of thinking like a city is supporting city change makers that are working outside the status quo, and are trying to make a very different future begin in the here and now. Researchers call these grassroots niche innovations (Seyfang & Smith 2007), where change makers develop niche ideas and activities that challenge mainstream orthodoxy and can become the big solutions of tomorrow. Creating spaces and providing resources for this kind of transformational innovation to bubble up is essential to the theory of change in this book. As I explore later, it can create solutions that are more alive to the challenges they face, long lasting and accepted by the communities they are embedded in.

Crucially, this approach to change is also about being able to control, evaluate and filter the flow of ideas and resources that shape places. The Uruguayan political theorist Raul Zibechi (2012) has talked about how particular places need to show resistance and stop certain flows that harm them. For example, too much global finance flowing into a place through speculative development, overseas foreign investment and global supply chains can set up an extractive economy where money and ideas are sucked out of a place. Letting wealth disparities go unchecked creates precarious jobs as the corporate controlled gig economy grows to service higher income groups. Supporting the extraction of fossil fuels and other natural resources accelerates climate and ecological breakdown.

Finally, thinking like a city means taking a systems approach. There are several elements to this. First, this requires an approach based on holism and interconnectedness, where a city can be regarded as a system made up from several subsystems such as housing, water, energy and food which interact in complex ways but come together, intentionally or otherwise, to create the things that we rely on for life (Lehmann 2015). These subsystems are created through a range of features including investment, planning, institutional and financial decisions as well as social norms and our natural world. They come together to create very mundane things like water from our taps, electricity in our homes or food in the grocery store. These subsystems combine as a "system of systems" roughly working together and producing the things we need in our everyday lives. Researchers have called these provisioning systems (Gough 2019) – the basic elements that satisfy our needs and underpin our well-being. The challenge going forward is how we arrange these provisioning systems in a way that supports human well-being and within the boundaries of the natural world (Vogel *et al.* 2021)

Therefore, while our work might focus on one aspect of city life, we have to be aware that it is part of a bigger urban system that shapes and influences it. To capture this breadth, sustainability researchers often refer to transformations across socio-technical and socio-ecological systems (Geels 2019). What they are highlighting is that system thinking encourages us to consider the different elements that combine to make cities – across politics, economics, society, finance, culture, behaviour change, institutions, technology and ecology (Foxon 2011). Our ultimate aim is system change in a city across all these areas, structurally changing the way the city operates, which in turn will create the conditions for change in other areas. When activists say they want system change, this is what they mean. They do not want small adjustments, they want longer lasting structural change across politics, economics and social issues that become hardwired into the way cities function.

Second, systems have unique and complex properties that we need to be aware of (Homer-Dixon 2010). They tend to be more circular rather than linear, with iterative feedback loops between system elements. This system feedback can be both positive and negative, setting off spirals of innovation and regeneration, as well as collapse and decay. They also include interactions between human and natural phenomena, so for this reason we need holistic practitioners who are alive to different types of system properties. There is also an element of what is known as emergence, where a system can generate outcomes and effects that were not previously anticipated. But systems are also adaptive; they are full of people with knowledge and resources who can manage and adapt to change. Ultimately, one of our tasks is to build resilience in our city systems so they can both withstand and survive current and future shocks and disruptive events, but also change the nature of the city so those shocks are reduced in the first place (Coaffee & Lee 2016).

Third, we need to be alive to the different levels at which systems operate. Multilevel perspective (MLP) researchers point to three: the larger landscape where longer term trends unfold normally over decades; the regime where formal actors create policies and regulations over the medium term, which can be considered the rules of the game; and the niche level where grassroots innovators constantly introduce new ways of doing, working and thinking which can disrupt and influence the regime (Geels 2010). Ultimately, how these levels interact is important for making and managing transitions from a less to a more favourable state (Bulkeley & Newell 2015). This is our key task in saving the city: supporting niche innovators, influencing regime rules and actors while also managing whatever landscape changes come our way.

Taking the example of housing we can see what this means in practice. Housing, or at least the right to shelter, is a basic human need. But houses do not simply appear. They are brought together through a complex set of actions in a housing system. Land has to be bought and assembled, design work undertaken, trades people skilled up and deployed, the planning system enabled, materials made and delivered, marketing created to attract residents, finance arranged by the developer and buyer, legal arrangements and tenure agreed between those involved in the home. A vast range of people, networks and resources are put in train, most of which span far beyond the immediate vicinity of the home being built.

The key aspect for our approach is how to intervene and change the way this system functions to better respond to the triple emergencies and build the conditions for a flourishing, safe life. Changing the housing system would include, for example, localizing land ownership, shifting planning towards community needs, restricting corporate developers, creating sustainable skills, improving design, ensuring affordable finance, communalizing tenure, restoring biodiversity and radically reducing the carbon intensity of materials and home energy demand. Changes need to happen, then, across a complex range of system elements. Changing one element is a useful start, as long as we recognize ultimately our task is to change housing as a system.

Act

The second strategy area is action. Being in emergency mode means learning what challenges we face, but also how to act. Knowing about the problem is the essential first step. But knowing too much without knowing what to do can induce a sense of powerlessness and even panic. Paraphrasing an often-used quote, if you are going to shout fire in a crowded theatre, you have to also point out the exits. Developing appropriate solutions and options for action can empower and offer focus. As first responders in the face of converging and accelerating emergencies, what knowledge and skills do we need so we can support, intervene and improve without falling back on panic, blame and confusion? The next sections introduce three strategic areas.

Go upstream

We need to think and learn to prepare for action. The further question is what can we do, especially in the face of such overwhelming, complex and global issues? Inaction usually emerges because we feel overwhelmed, powerless, confused, isolated, angry or even a deep sense of sadness and loss. But in an emergency situation having a clear sense of what we are up against and what will work is crucial. The good news is that there is lots we all can do that will help support positive change. It starts with locating ourselves, and our actions, further upstream.

Think of your actions placed along the course of a river flowing from the mountain to the sea. Some actions are more upstream, others more downstream. Upstream is where the problems originate and further downstream is where people are largely dealing with the fallout of the problem. There is further complexity: new problems may be added on the journey downstream, established problems grow, become more entrenched and multi-dimensional, while others might be resolved.

What is noticeable about much action on climate breakdown, nature loss or social inequality is that it takes place downstream in our daily lives, where the problems are already known, established and significant. Once the "cat is out of the bag" there is less we can do to stop harm being done to us, others and the environment. When we encounter these problems downstream in our daily lives they can feel overwhelming. Think about poor air quality, the lack of affordable housing, or the daily experiences of gender or racial inequality. In response, a range of bite-sized actions are often proposed that are more manageable and doable. These can include actions like recycling or changing our behaviour as consumers.

This focus on smaller downstream actions is reinforced by a range of well-meaning actors and organizations across local government, charities and think tanks. This approach has recently been popularized by the incredibly influential nudge theory first developed by behavioural economist Richard Thaler (Thaler & Sunstein 2008). While the idea of nudging people in small steps to adapt their behaviour is largely sound, it is part of a broader trend towards individualizing and depoliticizing how we approach and understand the big issues we face in our lives. The unfortunate reality is that small actions lead to small change. It is worth taking a closer look at some of the limits of this approach.

It is understandable that we focus on these bitesize actions. There is only a finite number of hours in a day, and even less that people are prepared to commit to deep social action. Few of us consider ourselves full-time activists. Most of us are tied up just getting through the day, finding time for families, friends and work. But when we focus on many smaller downstream actions it reduces the time we have to go upstream and explore and understand some of the more structural causes and where and how problems originated, especially historically and politically. Think about the time we commit to figuring out whether various plastics are recyclable rather than lobbying for less packaging or demanding that the local state use tax money to recycle more, asking people not to idle in their cars rather than demanding that the motor industry switch to electric, or trying to find lower impact plant-based food products rather than campaigning against deforestation. To paraphrase, "an ounce of upstream prevention is worth a pound of downstream cure".

The key task is to connect our downstream actions with what is happening upstream. On their own, downstream actions often pit individual against individual in a battle of who can do the right thing. The issue is that almost none of us can do the right thing in such a compromised and out of control world, where large, powerful and globalized forces structure our daily lives. Clearly, it is not just an either-or situation. Bitesize downstream actions are useful, and they do need to happen. They play a role in supporting our well-being, making us feel useful, creating an ethical approach to what a good life is, and social cues to others around us that we are doing the right thing. If they allow us to feel good about ourselves and others then we should do them. While they will never be enough on their own, they are part of what is called "prefigurative politics" (Maeckelbergh 2011), where we literally act like the change we want to see.

To go upstream, ultimately we need to act on those who hold money and power, allocate resources and make policy and decisions. This is an approach that links everyday life to the bigger picture of corporate and government power and policy, encouraging individuals to see their actions as part of a larger change story. A particular focus should be policies, resources and organizations that fund, promote, maintain and extract fossil fuel and mineral resources who are responsible for a significant proportion of global heating and biodiversity loss. There are pinch points here as I explore later, especially around shareholder and pension fund action.

Let's take an example where action is urgently needed: city mobility. Action is largely framed in terms of the transition from the internal combustion engine to electric vehicles. This is a positive move. It will reduce carbon emissions, tailpipe and noise pollution, and might even bring benefits around road injuries. This offers those with the financial means to buy into the simple and relatively straightforward downstream action of upgrading their old car to an electric one. Problem solved? Not quite. What if we go upstream? While getting rid of fossil fuel automobiles is essential, replacing them with electric vehicles creates new problems and entrenches old ones. They maintain a significant demand for raw materials and harmful extractive industries worldwide, especially in terms of batteries, and increase demand for renewable energy beyond our current ability to supply it. Moreover, they do little to address issues of public health, biodiversity loss, the decline of street culture, unequal access to car use, road deaths, car-based household debt and the dominance of a small number of global vehicle corporations.

A wholescale shift from fossil fuel to electric vehicles does not meet the criteria of emergency action. Instead of time spent researching, investing and managing a new EV, going upstream would mean reorganizing our lives to be less vehicle dependent, to lobby our local politicians for better walking and cycling routes, or affordable and integrated public transport, organizing walking buses or car pools at your local school or workplace or joining a group that is campaigning on sustainable transport. If we also include neighbourhood car sharing or doing errands for people locally, we can also build community and address inequality. At the same time this decouples us from the giant corporate car and oil industries who are at the heart of the emergencies we face. Going upstream is difficult. It is literally swimming against the tide. But unless we anchor our downstream actions within bigger upstream interventions we will not get the emergency change at the scale and pace we need.

Find points of intervention

Now we reach a point where we find out there is much to do, and in a limited time period. This is a positive realization, as there are a range of actions to suit different abilities, experiences and circumstances. We can all take part in some way, in changing our city. The key thing is to connect with others and see it as part of a bigger, interconnected change game. When we think about change, we have to lift our gaze from what is in front of us – the diesel car polluting the streets, the poorly insulated home wasting heat, the supermarket selling over-packaged unsustainable meat, the developer digging up pristine land. These are single moments that cannot be ignored and require urgent attention, but in our broader change game we need to use them to motivate us strategically about how and where to act.

The Center for Story-based Strategy (2020) in the United States offers one of the clearest insights into this range of actions we can take on the issues we encounter. Building on the work of renowned environmentalist Donella Meadows (1999) on leverage points, they call these action areas "points of intervention", defining them as "a place in a system – physical system or a system of ideas – where action can be taken to interfere with the story in order to change it" (see Figure 3.2).

Taking the example of the global automotive industry, we can explore what each point of intervention means in practice – lifting our gaze from an actual car in front of us to the broader system that creates complex impacts at a planetary level. First, the "point of destruction" would spotlight the extractive industries – where resources are mined and processed – to make our vehicle. This is largely done by global corporations who not only extract wealth from lower-income countries but are responsible for poor working conditions, workplace deaths and local pollution. For a typical car the list of resources is huge – metals including

69

steel, aluminium, copper, lead, magnesium and titanium, electrical components and microchips, plastics, glass, rubber, fibreglass and various woven fibres. A vast global corporate supply chain comes into play to bring these resources together from mines, clear-cut land, or oil wells across the world. We can also add here the scrap industry where millions of car units end up after their useful, and usually too short, life, polluting and despoiling the environment, as well as the air pollution and associated deaths and chronic illnesses they create during their lifetime. A range of actions can be applied here, from direct physical presence at the places of resource extraction, raising awareness about these harms, support and advocacy with those living nearby, putting pressure on those responsible for local pollution hotspots, lobbying for cleaner air standards.

| Point of Production | Point of Consumption | Point of Destruction | Point of Decision | Point of Assumption |

Figure 3.2   Points of intervention
Source: Angus Maguire, Centre for Story-based Strategy.

Second is the "point of production", where things are made. The most important aspect here is the power of labour: the workforce. How can we engage with those who produce things to change what and how things are made? This is complex and dispersed. Think of all the individual aspects that combine to produce a vehicle – the assembly plant, components and parts, logistics and distribution, land, sea, air and digital infrastructures. There are car manufacturers, with the biggest five (GM, Ford, Daimler, Toyota, VW) massing revenues of over $1 trillion dollars combined. There is also the global oil industry, which continues to be a major impulse behind the growth of vehicles, with road transportation accounting for nearly half of all demand for oil in OECD countries in 2020 (Statista 2020). This production element is made up of a vast global network. Directly making almost 100 million vehicle units every year, plus the after sales and repair trades, some estimates suggest it directly employs around 14 million people across the world, with a figure four times that in associated activity (ILO 2021).

Given that the point of production is where people work, our actions in this area need to be sensitive to the fact that this sphere is where livelihoods are created, workers deploy their skills and abilities, and a sense of self and community are reinforced. Our actions are not about stopping production but arguing for production that allows people to thrive within the safe limits of our natural world. Engaging with trade unions is a key intervention tool, to understand what is happening in each workplace, build trust and understand the concerns of workers. It is important that responses to the converging emergencies do not appear anti-jobs. Our example of the car industry is full of highly skilled workers who will be vital in underpinning the emerging zero-carbon economy. Our emergency interventions need to support momentum for a transition away from dirty and precarious work, to good and green jobs, which are better paid and less at risk of automation (Valero *et al.* 2021). This is best done through a job-to-job transition to make sure no-one is left without work. We cannot simply make vague promises to those currently in dirty jobs that new work will appear: it will take government intervention. Trade unions already have a track record of leading the way here. In the 1970s, workers in the aerospace sector created the Lucas Plan, which highlighted how manufacturing plants could be converted to useful civilian production (Wainwright & Elliot 1981), while the Builders Labourers Federation in 1970s Australia undertook green bans and refused to work on projects that damaged parkland or low-income communities (Burgmann 2002).

We can also focus on who owns the point of production. As I explore below, we need to make the case for more diverse patterns of ownership and control in the economy, reducing the emphasis on large global corporate entities through a resurgence of more diverse city- and community-based enterprises. Greater community and employee representation and accountability is key on the way to this. This is a significant challenge in our example of the automotive industry given its alignment with large global corporate entities and the fossil fuel industry. Actions could involve organizing trade unions, workplace meetings, leafleting and showcasing alternatives.

Third is the "point of consumption". For many people this is one of the most familiar sites of action. It is what is most visible and known: the pristine car show room, the second-hand dealer, or online vehicle trader. Taking the argument to consumers is difficult; what we consume is often closely connected to our sense of self and status. Cities have been so closely aligned to car culture that motorized vehicles now play a central role in simply maintaining our daily lives. Our actions need to highlight the harm they do to us, our communities and our planet, but also the benefits of the alternatives: walking, electric bikes and affordable public transport. Many actions are relevant here including leafleting, showroom demonstrations, talking to customers, consumer boycotts and street theatre.

Fourth, is the "point of decision": challenging decision making and decision makers. This is an often less visible area of activity compared to mines, factories or retail malls. Millions of people support the global automotive industry by devising policy, securing finance, creating laws, allocating resources and land, developing advertising, investing in new markets, planning highways and designing car-based developments. Global financiers search the world for new investment opportunities to extract fossil fuels, invest in car production capacity or local car-based retail and housing. The good news is that these are all areas where we can deploy action. These range from direct lobbying, joining public committees and forums, opposing planning applications and taking shareholder action, supporting politicians who share our goals, or even standing for political office where they do not.

The final area is the "point of assumption" where underlying beliefs or assumptions about our action area are formed and spread. This is one of the most exciting areas as it focuses on empowerment, creativity and celebration. In an area as pervasive as the global automotive industry, action can be very broad. Car culture is an example of a belief system that is very pervasive. Getting around in a vehicle has become so normalized as an assumption that it can be challenged almost anywhere. Mundane acts like getting to work, buying food, taking leisure, visiting family and friends are all tied with using motorized vehicles. In many contexts, to do otherwise seems odd or even a sign of poverty and backwardness.

Advertising plays a dominant role in normalizing assumptions about vehicle use. Advertising budgets for most vehicle manufacturers are huge, and are mainly used to perpetuate the assumption that the car is still the main route that offers freedom, status, social power, and even sexual prowess (Wollen & Kerr 2003). Actions in this area can be as creative as you are able. Subvertising, for example, creates alternative street adverts which can counter the strong pro-car messages in our public spaces. Public demonstration of alternatives can be a great way to show people that there is a different way to get about. These can include local walking tours, e-bike trials, better cycling and walking infrastructures, incentives to use public transport, as well as direct action events like Critical Mass and Reclaim the Streets.

One of the most exciting and effective events that attempt to normalize assumptions to different kinds of mobility are car-free days, where city streets are turned into places of encounter, party and exercise. These are spreading throughout the urban world with impressive examples emerging in some of the most polluted and car bound cities such as Bogota, Jakarta and Delhi. Experiments with free public transport in cities such as Tallin, Luxembourg and Dunkirk also challenge the assumption that the only way to get around is to pay for it (Papa 2020).

Ultimately, these five points of intervention work best together. They can powerfully reinforce each other by linking the mine, factory gates, retail mall, boardroom, street and television advert. Creating a sense of a joined-up movement between them can be powerful, and will give us options to move between and link actions. As I explore below, the key issue is to use these points of intervention to forge alliances across a new breakaway coalition that can start to take emergency action.

Create city wealth

Cities contain abundant assets and resources: public greenspaces, buildings, institutions, fields, woodlands, waterways, roads. We can consider these our stock of common assets that can be repurposed and put to work in a different way to create safer and more resilient city futures. In our emergency context to save the city is to save the city economy: to regain control of these resources and assets so they can be mobilized to save the city.

Specifically, this means embarking on a rapid process of community-led planning, ownership and wealth creation at a city level. We need to support communities and city sectors to rapidly build resilience, solidarity, care and capabilities that underpin sustainable, safe and thriving livelihoods. There are so many rich threads to thinking and action here. Community employment and wealth-generating initiatives have grown rapidly through local and municipally-owned organizations and co-operatives. The idea of community wealth building focuses on anchor institutions in a city that can promote a positive cycle of local spending by supporting employment and spending contracts, especially to locally-owned businesses in lower-income neighbourhoods. Gibson-Graham (2008) encourage us to see the modern economy not just through the prism of capitalism, but as diverse entities made up of all kinds of exchange including community businesses, family employment, barter and reciprocal exchange. Large parts of the economy can actually be considered a social or solidarity economy geared to meeting everyday needs (North & Scott-Cato 2017). Others have called this the foundational economy (Bowman & Froud 2014) or what Karl Polanyi in *The Great Transformation* (1944) called the substantive economy, suggesting the purpose of the economy is to create a social foundation which can meet basic human needs. Coote and Percy (2020) highlight how certain parts of the economy – energy, health, housing and transport – should be considered as universal basic services provided at an affordable level for all and outside the logic of private profit.

Why are these ideas important now? We have got into a fairly disempowered and disconnected situation, where cities have largely lost the ability to determine their own future. A new city economy rooted in community needs is an antidote to the heavily corporate, globalized, extractive and degenerative model of development that currently shapes our places. City wealth building aims to relocalize by building alternatives to the supply chains, procurement deals and investment contracts of the mega corporate world that disproportionately shape our cities.

The disbenefits of organizing our current city economies continue to mount up at an increasing pace. The big limit of our globalized world is how it now facilitates a hyper capitalist economy. Capitalism is a way of organizing our economy that tends to extract as much value as possible from things and activities. It does this by commodifying and enclosing them (Chatterton & Pusey 2020). Things that were once held in common are privatized, owned, given a value and sold back to us. This has brought more of us, and the things around us, into what we call the circuits of global capital: the networks and flows that are controlled and owned by multinational corporations. Here is the problem. Huge amounts of economic value that otherwise would stay in localities are literally sucked out through supply chains, employment and procurement contracts. What drives all this growth is vast and increasing levels of fossil fuel and raw material use by the highest income groups. These "luxury emissions" (Malm 2021) are a huge problem – carbon emissions from the world's higher income groups who are burning disproportionately more carbon to maintain their lifestyle and consumption habits. It is also driving inequality in cities, with wealth increasingly concentrated at the top of the social pyramid in particular social groups and neighbourhoods.

While our globalized world and large corporate entities have brought benefits we need a new balance. We have created supply chains, as well as knowledge, media and communication flows that offer the possibility of instant goods and services from across our diverse planet. A large minority enjoy the benefits of consumer durables and leisure and work options that have enriched lives in ways that would have been unimaginable a few decades ago. Global connection has also allowed ideas, culture, music and politics to connect and spread. We need to make sure we do not reject everything out of hand. There are many elements of our globalized world that we need to nurture and protect, especially around understanding, tolerance and the spread of solutions and innovations that will support and supercharge our emergency response. We need to find ways that the monetary value from this global flow of goods, people, ideas and services accumulates within city limits rather than being externally extracted. We need to harness growth, employment and investment potential in new directions, to make our city economies socially rather than just economically useful.

A typical day for a moderate-income family can illuminate many aspects of urban life, which are shaped to unintentionally maintain this corporate-controlled, extractive city economy. You turn the morning radio on – to a commercial radio programme selling airtime to advertisers. Maybe you have kids watching Netflix or Disney – all part-owned by equity firms who also finance fossil fuel and extractive industries. You grab a cereal full of palm oil from deforested pristine rainforest, and boil water using gas brought to you direct from a pipeline through an area devastated by civil war and ecological catastrophe. You jump into your car, owned, financed and powered by some of the world's most powerful corporations and oil conglomerates, or you might grab a bus or tram that is owned by a multinational transport group. You find a space in a carpark owned by a predatory equity firm, head into an office built and financed by an aggressively expanding global developer, buy your morning coffee from one of the many corporate outlets employing staff below the minimum wage and using coffee from resource-depleted plantations, while your lunch is processed meat wrapped in single-use plastic and packed by low-paid non-unionized workers. In your break you find brief respite at a gym recently taken over by a global health giant, and on the way home grab a few things from a supermarket owned by one of the biggest retail corporations in the country, located on a retail park owned by a voracious property company which buys and sells land assets. You return to your home, recently built (badly) on former woodland and greenbelt, by one of the country's largest landowners and home builders, in turn financed by a global equity financier who trades land as a commodity. Grabbing a cab using the world's biggest peer-to-peer e-platform, you take your kids to the local swimming pool recently bought from the local state by a global private investment firm, then stop off at another retail park to go bowling in a global entertainment multinational, eating pizza made from ingredients distributed by freight from the other side of the country. The day ends as you check your email on your computer owned by the world's biggest computer manufacturer and cloud storage provider, and the kids read a book on their electronic tablet using their subscription from the world's biggest online retailer. You book a long-awaited holiday on a global online lettings platform, and a couple of essentials from the world's largest e-commerce multinational, before plugging in your electric car into a battery bought from the world's fastest-growing automotive and clean energy firm.

None of this is surprising and none of this is your fault. It is just the way our lives have been shaped by an expanding universe of multinational corporations, whose job it is to expand their market share, extract monetary value from daily life, and maximize profits for shareholders. This dense network of global corporate entities, facilitated by national governments, act in certain ways, all of which are usually bad for local communities. They develop supply chains and procurement contracts with other corporations, rather than smaller local providers and suppliers, they source goods based on cost rather than social or environmental benefit, they drive down wages in order to increase their profit margins, they extract raw materials that destroys biodiversity, and many lobby to reduce social and environmental standards to ensure there are fewer checks on profit growth. At every step in our mundane daily lives, we have unknowingly become tied up in the globalized activities of multinational corporations: fossil fuel and mineral extractions, low-paid labour and child exploitation, illegal employment practices, land conflicts and displacements, environmental pollution, accelerating carbon emissions, e-waste mountains.

The good news is it does not have to be this way. At each moment of our daily lives there are opportunities to do the exact opposite: create, capture and benefit from value that currently goes to multinational corporations, create local employment opportunities that guarantee well-paid jobs, restore the environment, build community accountability and democracy and respond to local needs. None of this will come easy, and will not happen simply because we want it to, or because those that can afford to do so make a few more ethical consumer decisions.

There is a whole new geography here. Rather than the conventional heavily zoned city model of centre and periphery where activity feeds a dominant centre, we need a more diverse, but deeply connected, mosaic of communities. The big corporate global economy relies on concentrated downtown centres, where wealth, land and building values can be concentrated and maximized. This creates a virtual spiral of commodification, privatization and profits for larger entities who control, own and gain from central areas. The antidote is to shift and de-emphasize the function and role of downtown central areas, creating more dispersed and community-based employment centres, but also reimagining downtown areas as civic hubs: car-free places of entertainment, arts, culture and politics as well as community-based trading.

We have to get organized. We literally need to take back control of our city economies. This response partly relies on finding ways to introduce new international legislation that can set limits on unregulated big capital. But it also requires local and national politicians as well as enlightened local business and civic leaders, prepared to back city wealth building as a new mission. The question of ownership is central here. Communities need new powers to control land and buildings in order to generate local employment hubs, incubate community businesses, grow food, produce energy, build community homes, reskill, and make a range of goods and services responsive to local needs. If their power is unleashed, communities can bring forward a host of workable solutions to our emergencies.

Build

We now arrive at our third action area: making and building. I want us to think broadly about building. Not just in the sense of building actual things, but also building momentum for a new way of thinking, acting and dealing with problems. Let's take a look at these.

Build a new story

Stories drive our life. They form and change ideas and beliefs and shape a sense of what is possible. They inspire, inform and contextualize. Offering a convincing story, as well as being a good storyteller, is an essential part of saving our city. But here is where we also face a problem. The current story of our cities is shaped around some core and entrenched myths which largely need debunking. The most compelling is, as the sociologist Harvey Molotch (1976) reflected on in the US at the height of the postwar development boom, that the city is a growth machine where economic competition is put before the well-being of people and nature. In this story of the city, economic growth, increased consumerism and urban expansion is always good. This has created a foundational myth that has taken hold about why cities exist: to position themselves as best they can in an ever-expanding and consumer-oriented corporate-controlled global economy. The story presented and sold to citizens is that competition with other places both near and far is beneficial. City boosterism, urban branding and marketing is big business, often focusing on claims to be the best, backed up by the use of city indicators and league tables focusing on quality of life (Wilson & Jonas 1999). Scratch the surface and such indicators are actually about which cites have the best suburbs for high-income groups and gentrified downtown centres.

The second myth is that the current configuration of power and politics is functioning well and can deal with the problems ahead. This is a significant risk given the weaknesses and limits now showing in terms of city governance as they face converging emergencies. Centralized city governments based around mayoral or chief executive offices certainly provide momentum and focus. But they tend to be unresponsive to the complex and hugely varied needs of large city populations, susceptible to power hoarding and vulnerable to lobbying from powerful groups. Centralized offices of power lack a more general ability to respond to intersectional identities. They also tend to perpetuate the power of the dominant culture especially around ableism and white, male privilege. Moreover, our political institutions are out of step with current levels of complexity. Most were designed in the nineteenth century through trade offs and conflicts between powerful social and landed interests and an emerging labour movement, when urban areas were smaller with fewer enfranchised voters and faced less complex problems. Ask yourself, if you were to design a system today to run a city, how similar would it be to what exists?

What we have lost in our story of the city is what a city is actually for: to offer a good life to all, as well as welcoming new citizens and those in need of shelter or asylum. The consequences and costs of current city life are not built into the story, but they are the invisible backstory. These include the vast supplies of fossil fuels and raw materials that are mined and extracted to create our goods and services, the low-paid and precarious labour that supports everyday city life in the cleaning, health and hospitality sectors, the long-term poverty, dereliction and lack of work facing many city neighbourhoods, the slow degradation of local ecosystems and the unsafe accumulation of atmospheric carbon, chemical fertilisers, air pollution or plastic waste. The ecological and social footprint is simply not part of the story. We urgently need to build a new city story that can generate real material effects in changing political decisions, how people engage, how resources are allocated, and ultimately whether we can thrive and be safe.

Many enticing new stories, and storytellers are emerging. The Doughnut Economics Action Lab is a growing global community of city change makers who want to tell a new story about the future city. Based on the work of Kate Raworth (2017) they propose a safe and just space of the doughnut, and it has become one of the most compelling ideas for cities as they go forward. This doughnut economics approach weaves together two key ideas that allow us to tell a different city story. The first is the idea of a social floor that no one should be allowed to fall below. This floor guarantees everyone access to adequate basic services including health, water, housing, transport and education, as well as recognition and rights. The second is that there is an ecological ceiling which humanity should not exceed. Keeping life below this ceiling ensures we live within the safe limits of our natural world in terms of the carbon we emit, the chemicals we use, the water and land we draw on and the waste we generate. Our Earth is a relatively well balanced set of interconnected ecosystems and given all life depends on this, including the human species, we would do well to respect its boundaries. Between this ceiling and floor is a special place: a safe and just space for humans to thrive. We need to ensure we do not undershoot or overshoot, and find ways to operate and live well in this place. It is a great visual story for our times. Current work shows how far off most countries are in terms of this safe and just space of the doughnut. My own work with Climate Action Leeds shows how almost all aspect of life across social and ecological, as well as local and global dimensions, in my home city fall outside of the doughnut (Leeds Doughnut Coalition 2021).

Part of new city stories is changing what measurements we use to gauge progress. The crucial reframing asks: how can everyone, locally and globally, prosper while living within the real limits of a single planet? A key part of this journey is to think again about how and why we use gross domestic product (GDP) as the main measure for human progress. A constant upward trend for GDP has become an almost unquestioned mantra amongst policy makers and politicians. But it is simply an aggregate measure of activity in the economy. It does not differentiate between what harms or protects, what allows us to thrive or not. If goods and services increase regardless of what they are, GDP goes up. If more woodland is cleared for development, more polluting vehicles are in circulation, more patients in hospital, GDP still goes up.

We have to get smarter, and ask ourselves what is it that we want to measure human life against. The good news is that there are a range of measures that city makers can use right now that can support a different city story. The Thriving Places index (Saunter 2022) focuses on aspects beyond GDP that allow city residents to really thrive: healthcare, sustainable transport, affordable housing. The Social Progress Imperative creates a city dashboard and alerts city leaders to how well they are doing in three key areas: basic human needs, foundations of well-being and equality of opportunity (Haines 2022). The key task is to make these indicators the central aspect of what we measure. What really matters is the quality of our lives. A growing economy is only one part of that, and in our heavily commodified and privatized societies it often does more harm than good (Hickel 2021).

A new city story has to be built in a very different way. It does not come from the marketing hubris of city leaders, no matter how charismatic they are. A story that works makes sense to a wide cross-section of people. It reflects back their life, and importantly the life they would like to see. Crucially it also does not hide from the challenges ahead, but offers an honest appraisal of what needs to be done. It is a story that is coproduced for and by the people. This is essential if it is to be accepted as legitimate and authentic. Created well, a new city story will take time to build, and will be open to change. It will not just focus on fixed features, endpoints and goals, but will emphasize process, relationships and lived experience – what it means and feels like to live in a place, especially from the point of view of those living the most difficult and restrained lives. It will avoid the platitudes so common in contemporary city marketing that is fuelled by inter-city rivalry and a race to capture scarce external resources. Instead, new city stories need to envision a post-scarcity world: the vision that beyond private property and inequality there is an abundance of resources and assets that provide plenty for everyone, if managed and distributed fairly (Benanav 2022; Kropotkin 1974). Articulating how and why things have to change is a key part of building a broad social licence for the radical interventions needed. This is a "whole city story" which has cross-sector support. It is a story of the near future, that tells the truth, that stresses the positive outcomes for the majority of people, and which inspires the city's varied and diverse sectors to get into emergency mode as first responders. It is to this coalition of actors that I now turn.

Be part of a breakaway coalition

Who is going to do all this? The momentous task of turning the city on its head, bringing it back from the brink of collapse. This is a key strategic question and it is one where plenty of thinking has been undertaken. The short answer is, everyone, but not as you know it. Key to emergency action is building a breakaway coalition made up of the city's diverse sectors who are prepared to challenge the status quo and work together to do this.

Partnerships, alliances, groupings and coalitions are the stuff of city development. There is a long history of people coming together in cities to develop new ideas, create momentum and launch initiatives. Researchers have highlighted four key sectors that typically interact: the business sector, both large and small, the university and research sector, the public sector, and civil society which encompasses a whole range of charity, voluntary, community, not-for-profit groups as well as social movements. They all bring very different skills, abilities and experience to the table. They are often called the quadruple helix, reflecting how these four sectors interact almost like a corkscrew (Paskaleva, Evans & Watson 2021). They are the DNA of our cities, where people and groupings come together to share knowledge and address common problems.

This all sounds good. Who would not be for people working together? But there are two key issues to address. The first is power and representation. Our four city sectors are not equal in terms of status, power, reach and decision making. New public management scholars (Connell, Fawcett & Meagher 2009) have highlighted how the more dominant corporate and business sector often leverages its huge resources to set the agenda, influence policy and the allocation of resources. In this context, the other sectors fall into line to do their bidding. Cash-strapped public bodies have limited options and turn to and facilitate a pro-big business agenda, universities look for growth and investment opportunities to manage their over-leveraged budgets, service mounting debt and move up league tables, while civil society groups scrabble for crumbs, forced to play the same game of privatization, commodification and growth at any cost (Brenner & Theodore 2002). Some have suggested this situation represents a new kind of bio-power, where neoliberal market economics has become so accepted as the norm it can now discipline city inhabitants and govern invisibly and uncontested (Keil 2009).

The second is purpose. Up to now, city coalitions have played a role in innovation, whether it is improvements in how we deliver housing, promote mobility or protect ecology. But their focus has been to innovate within an established frame of what is politically acceptable, socially feasible and financially tolerable. As we go into emergency mode, we need something very different. The luxury of incremental change has passed. The challenge ahead is to move beyond coalitions that advocate for what is possible. Instead their task is to advocate for what currently seems impossible: rapid transformational change to reverse climate and nature breakdown which can be widely communicated and justified, while also increasing social justice as an outcome. This is a huge and complex task. It needs a new approach and rationale for coalition building and partnership working.

City actors in a range of contexts are turning to this task. We need people who are prepared to say, "I am no longer going to work in the old ways, they don't make sense or reflect the challenges I see". These are the breakaway actors. Those who are no longer prepared to use their day jobs, power, access to resources and worktime to hold up a system that is actively causing harm to people and planet, at home and abroad. The task is for these breakaway actors to meet and form a broader breakaway coalition and act as emergency first responders.

Erik Olin Wright (2010), pioneer of emancipatory sociology, clarified some of the tasks for this breakaway coalition through what he called three strategies of transformation. The first is to promote rupture, a clean break with what went before. In many ways the Covid-19 pandemic, while not a desirable model for change, represented this kind of rupture that can usher in new practices. The second is to work symbiotically, corroding and changing existing institutions from within. Changing the way institutions operate, how they allocate positions of power and resources, will be crucial as we go into emergency mode. The third is interstitial strategies, seeding and supporting pioneer examples and experiments that can develop very different ways of city life that meet the identified challenges. Interstitial literally refers to a place "in between" what we have now and where we need to get to. While these tend to be micro examples, they are places of experimentation where we can prefigure the future in the present, and undertake rapid learning. They are the R&D for our breakaway coalition. The most successful routes out of the present combines and embraces all these strategies. At different times, and in different contexts they will all have their usefulness and will work together to create a self-reinforcing momentum that can push against the status quo.

Breakaway coalitions have a particular relationship to power. They are comfortable with owning and building power, in order to give it away. This is power-with or power-together, rather than power-over (Partzsch 2017). This approach is central to eroding what we earlier called the "dominant culture". Breakaway actors need to cede power and resources to those who are outside this dominant culture to build a coalition that is essentially anti-racist, anti-capitalist, anti-ableist and anti-adultist. Decentring is an important task to diversify where power and resources operate (Appadurai 2001).

Trouble-shooting

Our final strategic step is to recognize, deal with and ultimately resolve pitfalls, dangers and problems. To recap, cities are complex entities, they are a network of systems that encompass diverse people, institutions, cultures, business practices, infrastructures, behaviours as well as natural and ecological features. These interact in multiple, diverse and often unpredictable ways. As we have explored, there are internal and external dynamics at play: money, power, people, goods and services all collide and flow in and through cities. No wonder that the task is difficult, and that problems emerge, multiply and stubbornly remain. In fact, our approach accepts that we cannot always know or manage the city, and to use this as a strength rather than a weakness. We need a strategic direction that embraces and harnesses complexity rather than controlling it. One of the realizations is that rather than having a fixed plan, we need an approach to planning that is both principled and values led but is flexible and open to change. Let's have a look at some key areas to address here.

The first involves confronting the contradictions at the heart of urban policy making – what is anecdotally called "cake-ism" (the idea that you can have your cake and eat it, simultaneously trying to achieve contradictory goals). Urban leaders and activists need to be clearer and more honest about the consequences of continued growth. That continued growth of, for example, global retail, poor quality volume-built housing estates, car mobility, or fossil fuel energy use is incompatible with the kinds of cuts in carbon emissions that are now recommended by the global scientific community. With such tight timescales there are fewer options for gradualist reforms. To save the city means introducing new aspects of city life, and it also means constraining others. The era of relaxed and considered change ceased to be an option some years ago. Urban leaders need to take control of the future story of the city, and be clearer that a new route to prosperity does exist but it requires a major change of course.

The second involves dead ends, especially those that rely on overly technological solutions. The current way that urban society is configured places significant emphasis and resources on the role of technological innovation in problem solving (Vacca 2020). In turn there is a high level of public expectation that technology will always play a role in ensuring continued human well-being no matter what challenges it faces. Technology will, and always has, played a role in improving human life, especially if we see technology as the whole range of more straightforward artefacts and devices that humans use to manage and improve their daily lives. But there is a growing reliance on complex technologies, especially around global data, artificial intelligence and digitally automated systems that control and dictate more and more aspects of our lives.

The key point is that technology is neither good nor bad; it depends on the social context in which it emerges and the kinds of social, political and economic relations and patterns of ownership that create it. Under current arrangements of our deeply corporate-dominated globalized world, technological innovation is overly focused on improvements in consumer convenience, financial innovation and new areas of growth led by transnational entities for a relatively small group of high-income earners. Our task is to promote technological innovation harnessed towards social needs rather than the private accumulation of wealth. In the early nineteenth century the Luddite machine breakers of northern England were right when they stated that they were not against machines, but machines hurtful to commonality. This is certainly a spirit we need to take forward into the present (Mueller 2021). Whether it is advanced AI, automated transport or consumer platforms, the task is to create a digital commons, owned by and for the public which recycles and shares the benefits of technological gains (Scholz & Schneider 2016). We will need the best that contemporary technological innovation can offer – in terms of food production, energy harvesting or geo-engineering – but harnessed to common ends.

Solutions to the multiple emergencies we face do not just involve new widgets and machines. Our emergency responses also need to focus on rediscovering that a good life is simply having and doing less (Hickel 2020). A reduction in demand is the often-overlooked solution to many of the problems we face. The most effective solutions also rely on redesigning our city systems through governance innovations and changes in behaviour to create more social equality. We can see this through mobility. Given the central role the global motor industry plays in our politics and econo-mies, shifting from the internal combustion engine to electric vehicles is largely seen as the solution to our mobility crisis. A shift solely to electric vehicles is not a panacea. It will deliver some air quality and carbon reduc-tion gains. But it will bring new problems associated with high material use and leave many existing ones such as a decline in public health, road casualties and deaths, status anxiety and stress, and an erosion of public street life unaddressed. As I explore later, emergency solutions involve redesigning our working and leisure lives in cities so we undertake less mobility, as well as offering public transport as a free public good.

A final pitfall to avoid is an over-reliance on numbers, data and targets. The idea of the "smart city" has taken central ground as a way of thinking about tackling future challenges (Mora et al. 2022). Emphasis is placed on city infrastructures, services and their residents all acting smarter as a result of access to digital services and platforms. There is evidence that this does yield improvements, especially in terms of, for example, integrated mobility, smarter home energy use, or advances in healthcare (Silva, Khan & Han 2018). But gains from digital technologies are usually not shared widely enough, and tend to improve the life of already digitally connected, higher income groups, leaving bigger issues of social inequality untouched (Green 2020). What is missing from the smart city is a sense of the quali-tative and emotional aspects of the lived experience of a transition to a safer world; what life continues to be like for those already facing struc-tural disadvantage and digital exclusion, and what is actually important to us. Ultimately, outcomes which do not lock-in climate justice are not viable city solutions, no matter how smart or sustainable they claim to be.

Moreover, we have to be wary of ambitious target setting, especially around the growth of climate emergency declarations and pledges. Set-ting ambitious targets is the easy part. The hard work follows in terms of matching pledges to clear action plans, resources and accountability frameworks. In the previous chapter we also saw the dilemmas and prob-lems between net and absolute zero as well as carbon neutrality and the issue of offsetting. In any case, it is almost impossible to determine carbon reductions in real time at a city scale. What we miss from this over-reliance on data are the networks of care, reciprocity and compassion that will bring us together to navigate the difficult years ahead.

* * *

So now we have our strategic approach: to learn the challenges and unlearn dominant culture, tell the truth and show solidarity; to create city wealth, focus on pinch points and link downstream actions with upstream causes; and build a new story and coalition that can make all this happen. We now turn to the actors that can make this happen: the educators, researchers, entrepreneurs, designers, architects, planners, politicians, activists, financiers, consumers and non-humans. All of these have a role to play especially in terms of the power, resources, insights and institutions they have at their disposal. But more importantly, it is about shifting into emergency mode as first responders in our decade of transformation to save the city.

4

Players: who will do it?

Cities are a sprawling mass of things: buildings, pipes, wires, pavements, parking lots, machines, roads, tunnels, bridges. This landscape of things is often what we focus on when we think "city". But alongside these bits and pieces of city life there is a dense web of living entities, both human and non-human, that make cities. It is the relations between these elements and how they come together that define city life. In our next step on our journey to save the city, I focus on what I call the players: the people who will put our strategies from the last chapter into practice, and make the emergency moves I outline in the next chapter. These players operate through dense and interconnected relations with others. How each player sees and shapes itself, how they relate to others, and how they use, mobilize and change our landscape of things, ultimately will determine our success. These players form our breakaway coalition, acting as first responders that will tackle our triple emergencies. It is a lot to ask. It means taking risks, feeling uncomfortable and standing up to powerful interests. There is a huge responsibility to change direction and take rapid, urgent and decisive action. But this is what is needed if we are to save the city.

I have intentionally developed a broad list of players. I want readers of this book to find their persona in this change game. But more importantly, it is crucial to understand that a broad range of actors make and remake cities. Given the complexity of urban life it is often quite difficult to identify who actually does what, who is in charge and who can actually make things happen. This is both a strength and a weakness. There is no single office or seat of power that can determine the future of cities. Cities are certainly powerful, in the sense that they are full of power, and we need to work hard to uncover who has this power, where it resides, how it is applied, and importantly how its focus can be disrupted and shared as we act in the emergency context in which we find ourselves.

We have to think about power in a relational way (Anderson 2017). Power does not just emanate from a single source. Think about power more as a relationship that is activated across and between a network of people and institutions. The key challenge for saving the city, is how we shift from "power-over" to "power-together" (Gaventa 2021), where existing networks of power are reconfigured, and new ones activated that disrupt the dominant centre and allow us to take emergency action.

We also have to think broadly about expertise and experts. Reputations, professional bodies and institutions have been built around expert knowledge. In many instances, expert knowledge is an essential asset in saving the city. Those with knowledge in medicine, engineering, information technology, construction, energy systems, sociology, politics and economics, for example, all have a role to play. But it is how we use this expert knowledge that is key. Knowledge has to inform not merely influence; experts have to support as well as lead. We also have to democratize decision making and knowledge, especially by creating an open-source knowledge commons that can be widely shared to create workable solutions (Hess & Ostrom 2006). Everyday "lived" experience plays a significant role here, acting as a lens to check and manage expert knowledge and to avoid overgeneralizations that rely on top-down interpretations (Butcher & Maclean 2018). What communities on the ground need is access to expert knowledge so they can then shape and apply it to their own unique context.

In the following sections I introduce a range of players that we will encounter and engage with in our quest to save the city. The specific nature of these players will vary immensely across social contexts. What I have presented here are starting points to stimulate thinking and action about the breadth and diversity of people involved in making and saving the city. In each section I provide information about how these players work, what needs to change, what role they can play in saving the city as well as specific ideas for emergency action.

These are not siloed players that do not interact. Connections and relationships are crucial between players to create an effective breakaway coalition. The aim is to create more flexible, interconnected and cross-cutting identities where people step outside traditional ways of acting and move across institutional boundaries. Ultimately, our task is to disrupt the business-as-usual model of ceaseless economic growth that is pushing our Earth systems beyond safe limits and causing so much harm to humans. We need disruptive ideas from disruptive people – those who refuse to be categorized, normalized and tamed by the world around them. But not in the pursuit of new business opportunities. Our disruption needs to be focused on system change to tackle our triple emergencies. There are also a range of players that will make the quest to save the city a rocky and contested road. Deniers, lobbyists and distorters manipulate, slow down and confuse. We have to accept that they exist and use our strategies to counter them. Let's take a look at some of the players that can support our emergency responses – you might be studying to be a planner, have political aspirations, work in a housing association, writing a school report or masters' thesis, or attempting to set up your own social enterprise. Whatever your position, you can start to get into emergency mode.

The scientist-researcher: creating the experimental city

The classic image of a scientist evokes people in lab coats standing in a white walled laboratory talking earnestly and peering into microscopes or closely watching a read out on a computer monitor. Similarly, the persona of a researcher evokes a detached neutral observer patiently looking at data or forensically analysing interviews and reports for deeper meaning. Both share a common trait of being one step removed from the actual context of societal problems, political influence and the daily hubbub of city life. The scientist and researcher both play crucial roles in creating what we call basic knowledge where people use their insight and curiosity to follow patterns and develop hypotheses that can lead to new understandings. These can create breakthroughs, which are used by others to build policy and resources. Allowing space, and allocating resources, for these kinds of activities is vital.

But in our emergency context, there is an important refinement to these players. Notions of objectivity and separateness are illusory: we are all deeply immersed in and influenced by our particular contexts, backgrounds and experiences (Karvonen & van Heur 2014). Most science and research is connected to institutional, governmental and increasingly corporate agendas, creating outcomes that do not often enough address the triple emergencies, and can perpetuate many of the problems we face.

Our quest to save the city is to create an experimental approach that is grounded in the challenges we face, goes beyond maintaining or tweaking the way things are, and rebalances power across sectors and social groups. This experimental city becomes a living laboratory for our scientists and researchers, identifying and responding to our converging emergencies, and coproducing solutions with citizens (Chambers *et al.* 2021). There are many precedents for this approach. The idea of citizen science aims to increase the public understanding of science, but more importantly involve people in it. Recent research I undertook with colleagues through a project called Leeds City Lab found that coproduction can support this experimental approach and build a local consensus for more radical solutions like 20-minute neighbourhoods, free public transport or using natural materials like straw for house building (Chatterton *et al.* 2018).

This experimental approach to the city is also part of a growing tradition of public scholarship and action research (Mitchell 2008). The academy, often seen as a detached and elitist institution, can be recast at the service of major societal problems where research agendas are collectively prioritized, and its resources are used to tackle our converging emergencies. This does not mean expert or fundamental research insights are not relevant. It means they are utilized and applied in a collaborative process of solution making that can be useful right now. In my own discipline, we use the idea of public geographies (Fuller 2008), which takes what we know to directly engage in and influence public debates be they on climate change, urban inequality, transport, housing or energy.

This more public approach to science and research needs a collective home. Over the last decade or so, there has been growing attention on public-facing places focused on engagement and coproduction. These have come in various forms – urban labs, coproduction labs, living labs (Evans & Karvonen 2014). Climate emergency centres are also emerging as a focus for citizen thinking and action. These spaces all share a concern to gather people together across sectors, and provide access to data, knowledge, skills, ideas and policy so they can better address current city problems. It is in such places that a range of grassroots and research endeavours can bring forward disruptive, open-source solutions and innovations that can put our strategic approach into action, learning and unlearning, developing new threads of resistance and building practical alternatives.

Ideas for Emergency Action

- Funders and governments can focus resources on supporting emergency research that links climate, society and nature.

- Coproduction methods can be used to empower and involve those impacted by research and make sure outcomes have greater buy-in.

- Buildings and assets can be offered to create centres for emergency solutions.

- Researchers can trial live solutions in community settings to test and learn what works.

- Coalitions of researchers and scientists can lobby for and demand resources for solutions that address our triple emergencies.

Check out: MIT's Senseable City Lab; University College London's Urban Lab; R-Urban community resilience project, Paris; Civic Square, Birmingham; Scientists for Extinction Rebellion.

The teacher-educator: relearning community

Most of us have experienced some kind of education, from the youngest kindergarteners, to school, college and university students through to life-long learners. Education happens in many settings – makeshift classrooms in remote agricultural communities, large urban state schools and elite private universities. While there are huge inequalities across the globe in terms of resources and experiences associated with education there is a common thread: a desire to provide an education that will prepare people for life and empower them to fulfil their potential. More crucially, critical independent thinking is central to being able to understand, and act in, our world. The educator Paulo Friere (1974), was clear that education is really about developing a critical consciousness to understand your context, recognize and name power, and act to counter it.

We need a radically different approach to education – one based on a deep awareness of our triple emergencies and a desire to rapidly bring forward zero-carbon and nature-inspired solutions that also generate greater social justice. A striking feature of the global youth strike movement was school and college students raising their voices to tell the adult world they were not happy with mainstream education on the climate – that they were not being taught the truth about the climate crisis, nor the appropriate solutions to it. We have to take this as a serious critique of the limits of modern education based on institutionalized schooling aimed at disciplining learners and shaping them as workers for the world of neoliberal, corporate dominated work (Illich 1971; Mitchell 2018). To save the city, then, radically different learning content and approaches are required: to both "unlearn" ways of thinking that perpetuate the status quo, and learn skills and insights that can develop practical solutions that are climate- and nature-friendly as well as socially just.

Practice-based education is central here where problems are presented and explored by learners and teachers together in real-world contexts. Topics can focus on the big challenges cities face such as how to grow local food, shift from car dependency, build resilient local renewable energy, create community and co-operative businesses or the skills of zero-carbon placemaking. Our cities and their communities become classrooms without walls (Ward & Fyson 1973). Street-based learning, mobile and pop-up classrooms that directly interface with and involve communities can immensely enrich the learning experience. In these spaces we need to co-create emergency programmes of learning and doing, geared towards the task of creating zero-carbon, socially-just and nature-inspired cities.

This approach to learning is also underpinned by a key ethical shift in how we see ourselves and our cities. Inspirational feminist, educator and activist, bell hooks (2003) stressed that the aim of education is "teaching community". What she means by this is that teaching can happen anywhere, through opportunities for conversations that challenge the taken-for-granted way that our world is structured. We have to urgently repair the damage done by decades of neoliberal capitalism that has left our communities threadbare, fragile, disconnected and drained by an economy that has shaped us as individual consumers and precarious workers. As we go into emergency mode, learning the skills and build-ing the infrastructures that support and bring diverse communities together will offer a lifeline to a safer future. This will involve a range of community-based projects that I explore later: community economies, car-free mobility, place making, nature-based solutions and new forms of civic democracy. At the same time, educators also need to guard against an increasing commodification of teaching where education is regarded as just another product that can be packaged and sold at the service of globalized, non-local corporate needs, or generating skills in areas such as marketing, advertising, mass tourism or global finance that do little to support human flourishing or future planetary safety.

Ideas for Emergency Action

- Compulsory courses on the triple emergencies can create foundational knowledge to guide action.

- In places where education is still based around economic growth, consumerism, extractivism and colonial mindsets, students and supportive teachers need to organise and demand change.

- Governments and funders can support freely available community and street-based learning.

- Educational institutions can set up affordable and community-based learning environments open to a broad range of learners and where solutions can be trialled.

Check out: Highlander Research and Education Center, Tennessee; Black Mountain College, Brecon, Wales; University of Barcelona's compulsory Climate Crisis module; Project Mushroom climate justice network; Centro Sociale Leoncavallo, Milan; Universidad Popular de las Madres de Plaza de Mayo (Mothers of the Plaza de Mayo Popular University), Buenos Aires.

The entrepreneur: building the new city economy

Change does not happen through good intentions. Saving the city also requires harnessing the immense talents of those who know how to use business practices for good, and bring new products and services to the market place. We are all familiar with the entrepreneur, an often self-assured go-getter and risk taker, who through good fortune or hard work has created an innovation that has resonated with many people, and perhaps also accumulated some significant profit along the way. The entrepreneur is a central player in our game because currently we tend to place great faith in the market to solve problems. There is nothing inherently wrong with this. Humans have thrived and prospered through market-based exchanges for millennia, usually based on local production, trading and work that meet community needs.

But today's entrepreneurial spirit often gets caught in the trap of growth for growth's sake, and money for money's sake. Rather than exploring what kind of economy actually underpins well-being, entrepreneurial activity gets pulled into the rapid expansion of the globalized corporate economy, converting socially useful ideas into investment opportunities for shareholders and equity firms, and at the same time pushing beyond the safe ecological limits of our finite planet. The central challenge is how we create entrepreneurial activity that can go into emergency mode and rapidly bring forward solutions that are both socially just and fully address the zero-carbon and nature recovery challenge. In contrast to the free market pro-growth economy, the task of the entrepreneur is to build a new city economy. In work I did with colleagues for the United Nations, we called this the "social and solidarity economy" due to its strong purpose: to generate socially useful activities and solidarity between groups that can create trust, understanding and well-being (see Dinerstein *et al.* 2019).

There are three key tasks here. First, entrepreneurs in this new city economy support collaborative, peer-to-peer exchange and flat working structures (Robertson 2015). Rather than centralized flows of command

and power, new economy entrepreneurs act in a network of equal peers. This doesn't just make work more efficient. There is a strong ethical intent treating each other as equals in terms of remuneration and decision making. Underpinning this new democratic impulse are worker co-operatives, employee-owned enterprises and community-led businesses. This collaborative spirit also values data that is open source, shareable, and can be put to social use and improved again and again. Platform co-operatives play a key role here, providing a counter to the corporate giants who control peer-to-peer consumer platforms. These co-operative alternatives ensure that the value generated from peer-to-peer exchanges are recycled back into communities of members rather than distant dot.com millionaires (Scholz & Schneider 2016).

Second, new economy entrepreneurs develop community-based production (Centre for Local Economic Strategies 2020). This has a specific role in localizing production and re-embedding markets in the neighbourhoods they serve. Bringing production and consumption closer together has multiple benefits in terms of reducing carbon emissions, increasing responsiveness to local needs and generating neighbourhood-based employment. Part of this is a focus on circular, regenerative activities, especially regarding nature as something to be nourished and sustained rather than objectified, extracted and commodified. Finally, there is a scepticism towards perpetual growth and expansion. Quality of work and connections are valued, sharing work more evenly. There is a strong ethical commitment to broader social goals of justice, equality and intersectionality. The new economy entrepreneur examines themselves critically in relation to the dominant culture, exploring how we address the deep structural inequalities around race, class and gender as part of their working lives. While many of these trends are still emerging, they point to essential directions of travel that all entrepreneurs can explore, embrace and advocate for in their quest to save the city.

Ideas for Emergency Action

- City governments can create dedicated city wealth-building strategies and incubation spaces.

- New economy entrepreneurs can articulate that they are in emergency mode to set them apart from those that are not.

- Investors, grant bodies and developers can create specific funding streams to support the local social and solidarity economy.

- Local anchor institutions can support new economy entrepreneurs through targeted procurement and employment contracts and support community and cooperative ownership.

Check out: Cooperation Jackson and Evergreen Cooperatives, Cleveland; Catalan Integral Cooperative; Community Wealth Building Strategies in Preston and North Ayrshire; the Camden Climate Municipal Investment Green Bond; Stir to Action's Selgars Mill and the Playground for a New Economy, Devon.

The city maker: the interconnected practitioner

A huge cast of people are involved in the day-to-day running of cities: officials, public sector workers, civil servants, statutory agents, elected representatives. One of the issues we face is the distance between everyday life and city hall with local politicians often seen as distant bureaucrats. Researchers looking at city transitions call them regime actors, implementing and upholding rules, regulations and policies that run our cities (Smith, Stirling & Berkhout 2005). Clearly, many of the tasks they undertake may seem bureaucratic but they are useful, if not life-saving. Refuse collections, street cleaning, environmental regulation, pollution control, child safety, education standards – these are all part of the many things our city officials do for us, in return for the taxes we pay. But there is a lingering sense that something is not quite right in city government. We need a new breed of city makers – politicians, planners and policy makers – who have

a key role to play in mobilizing our strategy, meeting the challenges out-
lined in this book so far, and putting into practice our emergency moves.

Walk through the corridors of your local city hall or council office and
you may come across portraits of city fathers and city founders, the great
and the good from a previous era when an identifiable municipal project
emerged to manage and direct rapidly growing cities. These city leaders
often responded to the challenges of their fast growing industrial city
and were motivated by important public issues of their time – a desire to
improve public health, expand education and greenspaces and life opportunities for swelling city populations and tackle poverty, squalor, crime
and disease. They were city makers in the literal sense, creating a cultural,
health and education infrastructure that underpin what we now know and
cherish as the modern city.

Some city makers were part of a socialist tradition that regarded the
urban condition as a route to a new utopia, exemplified by William Morris
and the broader socialist inspired and romanticist arts and craft movements (Goodway 2012), that rejected industrialism and embraced a more

community, grassroots and folk based politics. However, more broadly we can locate the early municipal project in the political and economic context that created it. While the growth of industrial cities was fuelled by innovation, a desire to congregate or change the world for the better, it was also created through enclosure, dispossession and violence. Early capitalism forced people from village life and its common lands and drove them in desperation into the cities as low-cost labourers. City makers, then, while guided by a sense of philanthropy and responsibility, were also driven by a modernist urge for command and control through grand projects and institutions of power, or were part of the landed and capitalist establishment who saw stable and healthy urban conditions as a way to increase economic output for local industry (Perelman 2000). These kinds of dynamics continue today at a global level, as people are driven from ancestral and communal lands into rapidly growing cities in the global south (Midnight Notes 1990).

This is where we encounter a saviour mentality at the heart of the dominant culture that has set the tone for what city making is about. City leaders have focused on creating favourable conditions for local business and political elites, industrialists and those benefiting from the legacies of colonialism to maintain and enlarge their power. Ultimately this has led to the growth of an industrial, fossil fuel dependent urban economy that has created many of the problems we face today, especially around nature loss, social inequality, and high levels of resource use and carbon emissions.

Fast forward to the present – and our complex, diverse world. The key question we face is whether city hall, designed in a different era, is still fit for purpose? There have certainly been progressive interventions with brave city leaders and officials attempting to strike out in new directions. The recent growth of city mayors and their offices across the world have given extra momentum to respond to our converging emergencies. Progressive city leaders past and present – Ada Calou in Barcelona, Guiseppe Sala in Milan, Jamie Lerner in Curitiba, Anne Hidalgo in Paris – have shown how cities can take a more sustainable, people-focused direction. Many larger cities have also experimented with democratic innovations such as elected assemblies, citizen juries, participatory budgeting and crowd-sourced manifestos. As we have already seen, thousands of municipalities across the world have declared a climate emergency. This is an important initial step in responding to the severity of current events. These declarations have gained ground and have become owned by cross-party coalitions and multi-sector city partnerships, many of which have gone on to create ambitious climate emergency action plans (Ruiz-Campillo, Broto & Westman 2021). But the hard work starts after this. City leaders need to have an open dialogue with citizens and stakeholders to carefully understand the transformative and urgent nature of the changes needed in different areas of city life such as food, energy, work and housing.

This brings us to a key question: given the emergencies we face what do city makers have to do and be? I suggest city makers need to recast themselves as interconnected practitioners at the centre of our breakaway coalition, who are prepared to listen, work with compassion, and heal the deep divides that have emerged in our cities. This is a profound shift that transcends the boundaries between traditional departments, institutions, and even between the human and non-human world. It is also about reconnecting the city, its neighbourhoods, functions and zones. These interconnected city leaders deploy key aspects of our strategy in terms of enabling breakaway coalitions and supporting a new city story.

The strategy in this book presents challenges for city makers and elected representatives. Given the depth of confusion and the level of convenience and quality of life that a small subset of the population now enjoys, it is extremely challenging for those seeking election to tell the uncomfortable truth that we have built cities for part of the city's population to have a terrific time at great social and ecological cost. But now that inequality of experience needs to come to an end, and rapidly. There are of course a range of widely shared benefits that will follow from making these changes, and advocating for this future will be one of the key tasks for our new city makers.

City makers need to start by rethinking the siloes of traditional policy making. The very top level of city government needs restructuring around our converging emergencies and associated priorities. Traditional departments need to be repurposed to more accurately reflect the system level and far-reaching changes needed. This task is especially important for planners, highways engineers and architects many of whom have adopted professional practices based on siloed and disconnected understandings of the world. For example, "housing" becomes "zero carbon placemaking"; "highways" become "sustainable mobility"; "parks and countryside" become "nature recovery and restoration"; "education" becomes "social justice and community" and "economy" becomes "wellbeing, regeneration and circularity". Direct democracy elements will become the beating heart of this structure: standing citizen assemblies and scrutiny panels, open digital platforms for crowd sourcing future and emerging ideas, recallable community delegates in a broadened committee structure working across departments.

This needs to become the planning process for our decade of transformation which starts now. New working practices and policy priorities will take time to embed. As a central part of the breakaway coalition, the aim of city leaders is to use the resources, regulations and powers they already have – to reclaim the power of the state (Cumbers 2016) – and reorient them to this emergency mode.

Ideas for Emergency Action

- City leaders need to place themselves front and centre of a breakaway coalition and create new department structures, visions, targets and indicators to measure progress to make this real.

- City emergency centres can be established to make this new intention visible.

- Priority tasks include plans for tackling the triple emergencies with year on year targets and action plans, and citizen led scrutiny panels.

- Frameworks based on doughnut economics can be used for city planning and data analysis.

- Citizen assemblies and open-source platforms can be used to scope and test emergency solutions.

Check out: Amsterdam's Doughnut Circular Strategy; Fearless Cities Network; Climate Emergency Centres Network; Bologna's Laboratory for Urban Commoning; Leeds' Climate Emergency Citizen Jury; Barcelona's Aurea Social Hub.

The social activist: doing everyday activism

There are those who are already fully engaged with the task of social change. These are our social change activists, dedicating large parts of their lives to a better world. There is no single definition of a social activist, or forms of organizing and tactics that they use. This player appears in many different guises and forms, emerging from a rich and diverse tradition of social struggle (Hawken 2018; Chesters & Welsh 2010). The social activist encompasses established political parties; extra-parliamentary radical activism, often associated with self-identity projects, causes or extra-local concerns around peace, ecology and spirituality; place-based community activism; peasant or indigenous activism in subsistence communities; tech-smart activists in global urban centres; and highly mobile and transnational activists working in global civil society. The social activist will work across the varying points of intervention we saw in Chapter 3. They are crucial players as they advocate for and put into practice some of the more rapid and far-reaching emergency actions.

In the context of the triple emergencies, social activism is multiplying in exciting ways across a range of areas including anti-austerity protests, the Occupy movement, Black Lives Matter, student-led protests against higher tuition fees, sweatshop labour, migrant rights, or anti-fossil fuel and -fracking activities, the SunRise Movement, Extinction Rebellion, Youth Strike, Fridays for the Future, the Green New Deal and Build Back Better movements. Anti-racist struggles are becoming more central, offering powerful insights into community organizing, challenging structural power and unlearning and decentring the dominant culture.

Social activism is not the preserve of progressive groups traditionally associated with left-of-centre politics. Devoting oneself to change is a position adopted across all political outlooks. Clearly, what I focus on in this book are the conditions for promoting social activism that can respond to our converging emergencies – radical carbon reduction and adaptation, nature recovery with a strong social justice agenda. Equally, social activists will be more effective if they can engage with, but not be consumed by, more formal party political cultures that have constrained more radical change in the first place.

What I stress for our social activist player is their role in doing everyday activism. To overcome both the passivity and distractions that are common in our daily lives, we have to find ways to make the everyday more activist, and activists more everyday. Activism can be for everyone, it is about building new relations rather than simply taking risky or illegal action. It is about reminding people that many of the things they already do count, and that these can be shared and scaled up. There are a number of key issues to navigate to realize our potential as everyday activists.

The first issue is to jettison the ideal of the romantic, militant figure dedicated to revolutionary change through visible struggle against an objective oppressor (Sparke 2008). For such a figure, there is always oppression to fight, a utopia to be recovered, personal glory to be achieved. While this is a difficult ideal to live up to, it also ignores the more everyday and community-based activism that is less individualized and masculinist. This militant version of activism creates social distance between the ordinary citizen and the social change specialist and throws up unhelpful binaries between activism and everyday life, as well as the powerful and the oppressed. To save the city, we need a view of social action in all its unromantic messiness. Social action is not the preserve of the militant or the avant-garde. Nor is it about pushing against a singular or stable centre that if toppled will offer victory. Everyday activism is strongest when it is collective, where solidarities and connections are forged between different groups. Everyday activists develop place-based initiatives and networks where these kinds of encounters can happen.

But everyday activism is more than local. It is deeply transnational (Sundberg 2007). Everyday activists will network and connect with people and struggles, weaving together bonds of solidarity and understanding, sharing stories of struggle and experiments with alternatives across borders. Fresh forms of internationalism emerged during the 1990s in response to the excesses of global neoliberal capitalism. Global networks such as the World Social Forum and People's Global Action brought together trade unionists, community groups, farmers and Indigenous leaders from across the world (Mertes 2003). These transnational alter-globalization movements built democratic self-organized networks of activists able to mobilize groups against what felt like the relentless onward march of the global growth economy.

A final issue for everyday activists is a shift towards professionalization, a trend which is becoming much more evident as the neoliberal economy penetrates further into our lives (Mayer 2009). Social change activities are no less susceptible to our business friendly, digitally savvy, corporate dominated world. Opportunities have emerged for activists to expand and connect through business models, online platforms and compelling branding. A social enterprise model has slowly crept into activism where change agents spend much of their time chasing grants and reporting on outcomes rather than doing social action and building relations. Today's activism can often be a route for career progression rather than challenging the logic of work. This is supported by a cadre of professionalized transnational activists, often backed by international agencies. While this professionalization has brought social change activities to a wider audience, it creates a distancing effect. It pushes to the edges more radical, local and grassroots activists who rely on volunteering, and importantly groups that are led by the most marginalized. It reinforces the division between professional activist and those they are fighting for.

Reduced time and space for everyday activism is part of a broader shift in our economy. As work and debt pressures spread more broadly, space and time for everyday social action are squeezed. When you struggle to make ends meet, you simply do not have time to think or act on the bigger picture. Those that traditionally supported activism, such as faith groups, trade unions and the voluntary sector, now also play smaller roles in terms of activism, and often reproduce business-friendly models and focus on income-generating activities. The language and business models of the social enterprise world now permeate social action, through the proliferation of grant opportunities, charitable foundations, awards and smart-tech solutions. The prospect of creating commercialized start-ups, products and employment opportunities as a result of everyday activism is understandably appealing for many activists in our precarious, debt-laden society. The potential of social activism that shakes the social consensus is further constrained through increasing policing and surveillance (Lewis & Evans 2014). Attempts to continually criminalize everyday activism, especially those using non-violent direct action, is a deeply worrying shift, and will deter people from voicing legitimate concerns about inaction through formal channels. Many people of colour already experience higher levels of harassment from police and part of everyday activism is showing solidarity with them.

So the everyday activist faces challenges: to stay true to ideals and values, finding space and time to do so, foregrounding the challenges and telling the truth in our distracted and distorted world, building meaningful activities locally while staying connected beyond the city, being radical enough to link action to evidence, but not so radical as to alienate people or attract the full force of the policing and legal system. The broader strategy in this book can support the everyday activist, especially if they see themselves as part of a broader coalition of city actors attempting to break away from the status quo. What is clear is that everyday activism will become more important as the need to stop the fossil fuel economy becomes more pressing (Malm 2021).

Ideas for Emergency Action

- Social activists can bring a campaigning approach to issues, focusing on how communities can win on issues relevant to them, and what resources, visions, values and aims we need.

- Centres for social action can be useful to normalize the role of activism and make it relevant to local struggles.

- Events focusing on learning and sharing the history of social struggles and what they have achieved can help a broader range of people see the usefulness of social activism.

- Explaining the need for non-violent direct action, and supporting those who take it, will become crucial. This will include active support for those affected by the police and criminal justice system.

- Activists can spend time understanding the lived experience of diverse groups in cities and this experience can be used to build our everyday activism.

Check out: Neon New Economy organisers, London; Stand with Standing Rock, North Dakota; Seeds for Change and Navigate Training Collectives; Centre for Story Based Strategy; Poverty Truth Commissions, UK; Tyre Extinguishers anti-SUV group, Scotland; Résistance à l'Agression Publicitaire (Resistance to Advertising Aggression), Paris.

The consumer: linking buying and making

At some point we all consume, even if it is for the basics of life. This is an area that we can all relate to, and where we can take emergency action. But we have to understand a bit more about how our economy works. Cities have always played a central role for trading, buying and selling. But as we explored earlier, more and more aspects of city economies are increasingly subject to commodification – creating goods and services that can be bought and sold. All commodities have a value: a use value that satisfies some kind of human need, and an exchange value that the owner benefits from when they trade that commodity. The difference between the cost of producing a commodity and the sale price creates surplus value. This understanding is one of the basic premises of Marx's political economy approach (Harvey 2017). Up to a certain point this is not too much of a problem. Exchanging commodities provides basic goods and services in city market places. It provides employment that underpins livelihoods.

But what defines our current era is commodification beyond socially useful and safe levels. Increasing amounts of surplus value are being accumulated by a smaller number of very large global corporations. Consumer markets are subject to monopolies where just a few transnational mega entities control price and how commodities are produced. Our deeply commodified world acts as a hugely effective device for large corporations to extract and accumulate ever increasing amounts of surplus value from our everyday consumer habits. What we also know is that this increase in commodities goes hand in hand with increasing resource use, social inequality and unsafe levels of global heating emissions. Moreover, previously non-commodified public goods are being brought into this circuit of globally controlled commodities. This includes areas that underpin our ability to thrive and be safe such as water, health, housing, transport and food. The circulation of money is now central to this commodity society (Holloway 2022). While we all need money to live, when money and financial transactions become the cornerstone of everything that defines our life in cities, we enter a future where financial and money markets determine our ability to have a safe and thriving future.

This commodity society only works if people see themselves as consumers in need of consumer goods. Based on the pioneering work of Edward Bernays (1928) over the last hundred years advertising has come to play a central role here, shaping and focusing our minds to want commodities. It is estimated that in the most advertising saturated environments, the average person might see up to 10,000 adverts in a typical day (Simpson 2017). It is the proliferation of sources that is most striking. Beyond the billboard and television commercial, adverts penetrate many more moments of our waking lives, facilitated largely by the dominant use of smartphones and portable devises. Adverts are also far more subtle now, hardly functioning as adverts, but quietly persuading, seducing and nudging us into consumer choices, or re-narrating to us how the corporate giants we are locked in to are actually the keystone of our well-being. Think about how Google and Apple now use strong threads of community and family in their marketing.

So how do we activate change in this consumer saturated economy? Our strategic approach provides some direction. First, it is probably obvious, but worth stating, that we all need consumer goods and services to ensure our well-being. The challenge is to explore how we meet our needs, without compromising the ability of others to meet their needs, or indeed harming the health of the planet both locally and globally. To address this challenge, our starting point is recognizing the huge disconnection between what we consume and how it is produced. We simply have no control over most of the things we consume. We act in an economy that is largely external to us and our communities in which goods are made from materials through complex and extensive supply chains, in production processes we do not know about, by people we do not have a connection with, or with finance we cannot control. In this context, entering into a blame game with other consumers about their choices does not make much sense. We are all largely stuck with a consumer-based society we did not make and in large part cannot control.

The task ahead for the consumer then is complex. There are some areas where changes can be made now. In middle- and high-income countries, there is a large segment of established consumers who can be incentivized to trim excess, for example, buying smaller cars, fewer gadgets, less processed food, fewer flights. This will need regulatory changes from government but also some of the system changes we have explored, such as changing debt, work and travel patterns. Beyond this group of mid-range consumers, the top 1 per cent of global earners consume far in excess of the other 99 per cent, and are responsible for what is called luxury carbon emissions from private jet travel, mega yachts, SUVs and even space travel (IEEP & Oxfam 2021). Stopping luxury consumption and the emissions that go with it is a priority. One of the most worrying global trends is the continued growth of what is called ultra-high-net-worth individuals (UHNWI), those with a net worth of $30 million. While they account for less than 0.003 per cent of the world's population, according to Knight Frank's annual Wealth Report (2021), they hold 13 per cent of the world's total wealth. Their activities are focused on global cities which offer opportunities for investment, wealth creation, ultra-luxury lifestyles and consumerism. This disparity is staggering and highlights the vast and growing inequalities that make up our world. Reducing the consumption habits of this group is a huge dilemma, for which there are no easy answers.

But we also have to change the structure of the economy. Consider an example of a large company who introduces a new product line, in this case a vegan sausage roll which becomes hugely popular alongside their already very popular line of meat sausage rolls. This seems like a really positive move that can address meat production, cruelty to animals, methane emissions from livestock and even perhaps deforestation. But the company does not reduce the production of meat sausage rolls as a result of the popularity of their vegan one. It simply becomes an opportunity for increased sales, commodity expansion and a branding exercise to expand its consumer base to more ethically minded consumers. To some consumers, it might feel like things are changing. But the same amount of deforestation, processed meat production and animal slaughter continues. In addition, there is an increase in global soya and battery-farmed eggs to provide meat replacement fillings, as well as the corporate supply chains that produce and distribute these. We have to be careful, therefore, about making false assumptions between the choices we make and whether they are part of structurally changing the economy. Unless it is on a very large and organized scale, consumers do not have the ability to send market signals through the economy to change production. Simply put, buying a non-plastic bamboo toothbrush does not provide a direct market signal to producers of plastic toothbrushes to reduce their production. In addition, any small gains in mature consumer societies are wiped out by the consumer habits of people in new, rapidly-growing markets in other parts of the world.

The bigger task, then, is beyond the store shelf. It is not about consuming this or that product. It is about changing what and how products are made, and importantly how we relate to our local economy. It is about a shift from being solely a consumer, to connecting with, getting closer to, and being involved in production. This is what Alvin Toffler called the "prosumer" in his book *The Third Wave* (1980) – where people start to produce the goods and services they would otherwise consume. The prosumer points to trends before the age of mass consumerism where the boundary between the consumer and the producer is blurred, and the local economy is based on a dense network of relations between traders and consumers. There are three key aspects to this.

First, the prosumer wants to reduce the distance between consumption and production. No longer satisfied with choice as a primary motive, the prosumer wants other qualities in terms of market relations. The main amongst these is shortening and localizing the distance between consumption and production. Much of this relies on seeking out, supporting and backing local employment generation, neighbourhood manufacture and community-led enterprises. Cost and value for money remain central. There is nothing inherent about localized markets that makes them more or less expensive. In fact, costs can be driven down as subsidies are shifted into city economies rather than global corporations. Shorter supply chains can reduce overall production costs, increase local employment, and give local consumers access to better quality goods and services. As I explore in Chapter 5, this is part of community wealth-building where money is respent locally creating a virtuous cycle of city-based prosperity. These efforts need expanding and connecting, ultimately replacing rather than sitting alongside more globalized and corporate-controlled avenues for meeting our daily needs. Clearly, these habits will not change the economy on their own. They need to be backed by an enabling local and national state that will make the prosumer the norm, not the fringe exception, especially by shifting vast state subsidies.

Second, the prosumer is also clear about the values they want to see realized through meeting their consumption needs. At times this will mean standing up for what is right, through for example boycotting consumer goods that have a direct link to harming people and the planet. Individual boycotts offer alignment with personal values. But to effectively send signals to the market to change how things are produced, they need to be scaled up and collectively articulated in many places by many people.

Third, prosumers simply consume less. Liberated from excess consumerism and associated travel time, prosumers will have time they can allocate to other tasks. Rather than deliberating on ethical consumer choices, our precious time is actually better spent on local action – supporting community businesses, spending more time with family and friends, volunteering at community projects, objecting to planning applications, or taking shareholder actions. This last one is a powerful everyday route to hold the activities of large corporations to account. Many of the giant global corporations that shape our lives are impenetrable in most respects, apart from the desires of the shareholders they ultimately serve. Groups of ordinary citizens have been getting together to acquire shares and using their collective voice to demand changes, whether divesting from fossil fuels, better pay and conditions for workers or respecting human rights. The final aspect about a market shaped by the prosumer is that it frees up vast amounts of city space that was previously devoted to the global economy. Retail malls, fulfilment centres and distribution warehouses, outposts of our global corporate consumer world, can all be repurposed for leisure, biodiversity and carbon sequestration, local employment, or city-based food and energy production, and all in a way that tackles local inequalities. So this is an exciting part of our quest that we can all take part in. As Gibson-Graham and their colleagues put it, it is part of taking back control of the economy (2013).

Ideas for Emergency Action

- The main action everyone can do in their role as consumer is simply to consume less, across all activity areas.

- Where consuming goods are still necessary for a good life, goods should be as local as possible, and support community and cooperative businesses, and employee ownership.

- Using the planning system and regulatory powers, city authorities need to incentivize and support locally produced goods that are as accessible and affordable as possible for local consumers, and restrict global supply chains.

- Direct action, boycotts and campaigning can be used to bring about rapid change where consumerism still supports ongoing and destructive activities, especially around the extraction of heavy metals and fossil fuels.

- Shareholder action and government lobbying can press for divestment from fossil fuels and improved labour standards.

Check out: Fairmondo online co-operative; Share Action responsible investment campaign, London; Library of Things/Buy Less centres, worldwide; Barter markets, Buenos Aires; Auckland's fossil fuel divestment declaration; Barclay's Bank boycott, UK.

The citizen: managing the urban commons

We are so often labelled through our roles as worker, teacher, student or consumer that we can overlook our role as citizens. The idea of citizenship is a powerful one. On a formal level it refers to a legal belonging to a particular place, and various responsibilities such as paying taxes or voting, as well as the rights and freedoms that come with that. It also has a more everyday meaning as someone who resides or dwells in a place. Being a citizen is also tied up with the modern nation state and there are many people who are denied citizenship for various reasons. This is one of our areas for action: how can we create cities of sanctuary (Darling 2010) which are open to everyone, both state based or stateless, and especially those seeking asylum or feeling persecution?

In spite of our deeply unequal world where some citizens clearly enjoy more freedoms than others, the idea of the citizen points to an important principle: that in return for various responsibilities we all should have an equal stake and voice in the future of our places. Cities still offer a common backdrop that we all can share. Even with the trends towards greater segregation, rich and poor residents inhabit the same city. This shared urban experience is thrown into stark relief in the face of an existential threat like climate breakdown. While the super-rich can protect themselves against social and natural vulnerabilities to a certain extent, there is a common fate to the emergencies that are unfolding.

In the context of this shared future, the role of the citizen is simple. With Andre Pusey, I explored how citizens can manage the urban commons for the common good (Chatterton & Pusey 2020). The "commons" is a powerful idea that plays a central role in supporting emergency action. It evokes a bounded common (the fields, the village greens and the forests) governed by those who depend upon it – the commoners. This use of the commons dates back to land enclosures in England, and the dispossession of peasants from that land, just before the onset of the industrial revolution in the mid eighteenth century. But it is a trend that continues right to the present day (Linebaugh 2014; De Angelis 2003).

The commons is such a powerful idea for our times because it refers to much more than land. It also encompasses physical attributes of air, water, soil and plants as well as socially reproduced goods such as knowledge, languages, codes and information. The commons is a political project that cuts across established ways of thinking about land and assets that is neither private nor public, neither capitalist nor state led. The commons refers to a way of reclaiming land and assets in a way that imagines ownership and governance as collectively owned and negotiated between and among populations. The commons can sustain our daily well-being in the face of the converging emergencies. They can facilitate new relationships between our players, which can, as we see in the next chapter, allow experimentation with our emergency strategy and moves.

There are three key stages here. First, we have to recognize that our cities are precious commons, and that our urban land and assets ultimately nourish us and sustain our well-being. Commoners recognize that current state and market-based approaches have led to deeply unequal ways of owning land and resources that no longer guarantee our well-being. Beyond enclosure and private ownership, there is plenty for everyone. We do not have to live with false scarcity imposed by private ownership. With the right policies, taxes and incentives we can create a new age of plenty (Coote 2017). The task becomes how to best manage this new city abundance for the benefit of all.

This brings us to our second stage, protecting and defending the commons. The citizen of the twenty-first century will only thrive on the commons if they are defended from larger private interests who have seen the commons as an opportunity for short-term gain over longer-term sustainability. Defending the commons will play out differently in different places – it may be protecting a local playing field or green space, a well-loved community or arts venue, or the right to access a free public service. In each case the key task is to mobilize the local political system, using whatever tactics and tools are available, which may mean using the legal and planning system, launching media savvy campaigns or the hard slog of local community organizing. At times this might also mean using civil disobedience and direct action, literally using our bodies and freedoms to protect our common good (Berglund & Schmidt 2020).

The final realization is that living "in common" means new skills and ways of relating for citizens. We have to demand that the truth be told, and force those who hold positions of power to listen, understand and act accordingly. If we are to live in common with each other we have to speak authentically to our fellow commoners about what we feel we are up against, articulate a strong set of demands that will fundamentally change the way cities are governed, maintain sustained pressure until these demands are implemented, and then set up scrutiny measures to make sure they are kept to. New skills and competencies in managing the commons will be needed and these will range from group facilitation and non-violent communication to participatory budgeting, open-source policy making and citizen assemblies.

Ideas for Emergency Action

- Cities can introduce new legislation to acknowledge, extend and protect common land and resources through an official register of common assets.

- Common ownership of land, housing, energy and food systems can be promoted through the planning and tax systems.

- Private and public services, especially transport can be subject to community-control and scrutiny.

- City wide cooperative agencies can be set up to support community and employee ownership.

- Supporting new arrivals and those seeking asylum can support a broader sense of the city as a commons for all.

Check out: Shared Assets think tank, UK; City of Sanctuary network, worldwide; the Land Workers Alliance, worldwide; Lilac Grove's Mutual Home Ownership Society, Leeds; Union Taxis, Colorado; HomeBaked community Land Trust, Liverpool; Wards Corner Community Plan, London.

Big business: the emergency response corporation

So, what is the role of big business in all this? One version goes that they are the heart of all that is wrong – destroying the planet, paying low wages, trampling on local communities, extracting wealth, and paying huge salaries and bonuses to bosses (Harvey 2013). Another version sees them as a route to all that is good – providing jobs, economic prosperity, generating investment, and meeting unfulfilled consumer demand (Kwar-tang *et al.* 2012). Reality always sits messily amongst these characteristics of hero and villain. Without a doubt, there are mega-sized transnational corporations, as well as their financial backers, who have grown too big and bad. They shape and direct our lives, habits, and even thoughts, in ways we do not even recognize anymore. They continue to invest in harm-ful activities, such as the extraction of natural resources, that accelerate dangerous climate heating and untold global poverty. Many function through human rights abuses, corruption, lobbying, child labour and tax avoidance. All these issues have to be dealt with as we save the city.

The world's largest corporations, their harmful practices, their bil-lionaire owners and financial backers will not simply disappear because progressive politics wants them to. It is more likely that their influence will continue to grow, as wealth and power is consolidated further through mergers and acquisitions amongst a handful of centi-billionaires and asset management companies. In all likelihood, they cannot, or do not want to, play a role in saving the city. Or at least in the terms laid out in this book. And even if they did, there would be so much greenwash, lobbying and confusion, how would we even know? What we do know is that big business is increasingly playing a role alongside city and national governments, providing insights, policy direction and management functions. As I explored with colleagues (Birch & Mykhnenko 2010), the broader doctrine of *laissez-faire* neoliberal capitalism based around promoting globalized corporate interests, commodity speculation, privatization and the fast movement of capital has become firmly entrenched as a way to manage cities. This shift has gone beyond a mere set of policy priorities – the whole distinction between city policy and corporate culture begins to blur.

We have to acknowledge these broader trends and focus on what we can change. What if some modern-day business corporations could double-down on the converging emergencies and lead the way to corporate reform? What if a new social purpose transformed them into the emergency response corporation charged with the immense task of promoting nature recovery, zero carbon and socially just cities? Again, this is not a "nice to have" addition. Radical change is also part of a self-interested survival strategy in the face of accelerating emergencies. For the corporate executive, their employees and stakeholders interested in saving the city, this means taking action in a number of specific areas.

First, it means a redefined mission. The great thing about corporations is that they provide a platform for a strong sense of identity and loyalty. This is a huge motivating factor for their employees and also those who consume their products. There is a deep trust and emotional connection. This is something to be used and nurtured as we tackle the triple emergencies. There is much to build on here, including Certified B Corps legally required to consider impacts on workers, customers, suppliers, community and environment (Marquis 2021), new strict Environmental, Social and Governance (ESG) accountability standards, and exciting new business-focused change networks like Race to Zero, 1% for the Planet, Realise Earth and Business Declares. The triple bottom line of people, planet and prosperity creates a starting point but we need to go further.

Rather than prosperity being an equal goal, the emergency response corporation sees prosperity as a shared outcome of nurturing people and restoring nature. We all need an economy where we can prosper, but it needs to exist within the limits of the planet and not at the expense of people (Jackson 2021). This change of emphasis is embedded in a more authentic relationship to customers, being honest about products and their impacts. Rebalancing profit-seeking amongst other goals is essential to move away from the "grow or die" mentality that has for so long shaped the corporate agenda. In fact, guaranteed future prosperity comes from nurturing and protecting what we actually depend upon: the Earth's finite stock of resources, and the broad ability of people to lead flourishing lives. Company profits and shareholder dividends are reduced to a minimum in return for a larger societal dividend that we all benefit from. The Doughnut Economics community (Sahan *et al.* 2022) frames this shift as changing the deep design and purpose of business especially in terms of how it is owned, governed and financed. Pioneering examples are emerging such as Faith in Nature, the natural beauty and health product firm, which has given nature a place on their board of directors, and the steward-ownership model used by the clothing firm Patagonia where all profits support the company's purpose rather than external shareholders.

Second, the emergency response corporation needs to see itself in an interdependent web of relations with other actors. This takes them beyond simple relations with consumers, investors and suppliers. Instead interconnectedness focuses on the circular economy where the corporation shifts from a simple linear throughput of resources and goods that are extracted, used and then disposed of (Bassens, Kębłowski & Lambert 2020). In a circular corporation, outputs become inputs in a closed loop that maximizes gain for people and minimizes impact on the planet. Certifications such as cradle to cradle (Braungart *et al.* 2007) are used to show commitment to this new circular economy. This is a fundamental shift in the production and consumption process. Not only are goods better made, they are not designed for built-in obsolescence, the short product life spans that are geared towards maximizing profit. Consumers become fixed-term leasers of goods that are then sent back for reuse, recycling or upcycling. This closed loop is an essential part of pulling back from the unsustainable levels of resource use and commodity circulation that humanity currently depends upon. While the loop between materials and consumer goods is closed, the loop also gets smaller – we simply consume less and bring them in line with everyday needs. In clothing, food, electronics, automobile or toy manufacture this is all possible. But there is no feasible closed or smaller loop fossil fuel economy. There simply needs to be a moratorium on these sectors, where coal, oil and gas become stranded assets, with international incentives used to leave them in the ground (Carter & McKenzie 2020).

The final aspect is the wider approach to organization and management. The emergency corporation operates in an increasingly complex, non-linear and fast-changing world. Its strategies have to reflect this. It rejects the older model of bureaucratic management, hierarchical executive leadership and profit maximization. Art of Hosting practitioners (2011) call this the "fifth organizational paradigm" where organizations adopt multiple strategies that all play different roles, but ultimately work together. These include the inner circle for trust and collective clarity and purpose, hierarchy based on project work teams, bureaucratic elements for essential tasks, and the network that scopes, connects and shares. Part of this organizational change is to learn from and mimic the natural world, its design patterns and organizational strategies. The emergency response corporation is "biomimetic": recognizing, learning from and mimicking patterns found in nature, and ultimately regenerating rather than degenerating the natural world (Pedersen Zari 2018). Not only does this create better products, but it can create more flourishing relations, understanding and care between people and the non-human world. Leadership here is much more collaborative, and aware of the interconnectedness and complexities of the human and natural world. The employee experience is also significant, in terms of pay, conditions, representation and workplace protections.

Ultimately, the corporation that goes into emergency mode has to coexist within the complex ecology of the city economy, alongside the public, community and independent sectors. There is no reason that they cannot find a new way to exist that focuses on city prosperity for all, higher quality and lower levels of affordable consumption, while also reducing material and energy dependency, promoting rapid carbon reduction, nature recovery and playing their role in greater social justice.

Ideas for Emergency Action

- The main area for corporate action is to get out of fossil fuels – in terms of research, investments, supply chains, products.

- B Corp and ESG reporting can implement radical changes in terms of procurement, supply chain conditions, travel, staff well-being, pay, energy and waste.

- Adopting zero-car policies and incentivizing active travel can support the broader relocalization of city life.

- GHG emissions can be identified, measured and actively reduced through zero-carbon reduction plans based on urgent and immediate timescales that also avoid offsetting.

- Committing to circular production and zero-waste strategies, and joining with others locally, can support a broader city-region circular economy.

- Shift to 100 per cent renewable energy and support local renewable energy production.

- Undertake compulsory triple emergency training for staff and suppliers.

- Employee ownership, staff representation councils, minimum and maximum salaries and four-day week programmes can all help the corporation address the broader social emergency of modern work.

Check out: Publix employee-owned supermarket, Florida; the Dansk Retursystem deposit and return bottle system; Business Declares an Emergency Network; Fairphone ethical mobile phones, worldwide; Mondragon Cooperative, Basque region.

The non-human: rewilding the city

There is one fundamental but overlooked reality in our quest to save the city. Cities are not just human entities. They are full of non-human lives that have their own ecologies, systems and patterns. My focus here is not the world of robots or cyborgs, but rather fully animate and sentient entities: the abundance of animal and plant species that make up our world. These are essential players in their own right, but also in terms of supporting and improving human life. The extent of this non-human living world in cities varies hugely, influenced primarily by the planet's varied climatic zones ranging across tropical, temperate or polar. But it also deeply reflects ongoing and historical and social processes: how nature is valued, how cultural, social, dietary and faith norms shape urban nature, especially in terms of activities like hunting, and whether current city form and function values and protects animals and plant life.

The human and non-human world has had a rocky relationship over the centuries. Since the scientific revolution of the seventeenth century, the natural world has been both externalized and inferiorized, something only valued in terms of its usefulness to the project of human progress (Lent 2021). The natural world has been hunted, pillaged, squeezed,

129

exploited, undervalued, ignored, belittled, to the extent that our fellow species now face extinction on a mass scale (IPBES 2019). Once a critical mass of extinction is surpassed, it will take millions of years to regain the abundance of life on Earth. This is the heart of the nature emergency. Humans live in a deeply interconnected web of co-dependency and survival with non-human species through activities such as food pollination, regulating temperature, cleaning air, creating soils, providing medicine and increasing well-being. Interacting with animals and natural environments, for example, has been shown to reduce stress-related hormones (Ewert & Chang 2018). Ground-breaking studies are beginning to find out that in forests and woodlands we breathe in phytoncides, chemicals that plants emit to provide protection from insects which also have beneficial effects on the human immune system (Li 2010). Forest bathing has emerged as an activity to reflect these beneficial effects. We are learning these crucial lessons too late, although it is clear that Indigenous peoples have been longstanding but overlooked advocates for this wisdom (Shiva 2020).

Cities, then, are often not seen as places for non-humans. Animals are tolerated, sculpted, contained, appearing in aesthetic or domesticated form, where they do not exceed numbers that challenge human life. But given the chance, as Covid-19 lockdowns showed, this non-human life can spring back and take a more central stage as the fast pace of the human species took a momentary backseat during the biggest "Anthropause" of recent times (Stokstad 2020). This non-human life cannot speak for itself, or at least in terms that we can easily understand. Our task is to comprehend our natural world and find ways to speak and advocate for them. What would trees, animals and plants say to urban planners and policy makers if we could readily hear them? What would they ask us to do differently? Progress has been made here, in terms of non-human rights, such as granting legal personhood to some animals and ecosystems (Cavalieri 2001). But we need to act much faster and bolder if we are to protect the web of life we depend on.

The task ahead is to rewild the city. Even the word "rewild" is loaded with meaning. Out of control, dangerous, unknown, exotic. But these are elements we will have to embrace to save the city. It is a risk, a leap into the unknown, a challenge to the arrogance and dominance of the idea of progress and of humans as the apex species. There are three elements worth considering here. First, we need to unleash non-human life in the city in profound and unknown ways. This is also a call to recognize what is already all around us, in our homes, gardens, parks and streets. Non-human life is everywhere. It flourishes in spite of us. Up to now, there has been an uneven patchwork of attempts to embrace nature: biodiversity corridors, bounded rewilding projects, blue-green infrastructure. As I explore in Chapter 5, we need to embed an urban design approach based on biophilia: a deep, rather than tokenistic, love of the natural, making space for it as equals, not as something decorative or beneficial to us. This is also about recognizing the life-giving and life-saving properties of these interventions. Letting nature back in will help solve some of the major challenges ahead, especially around the accelerating and deadly dangers of urban heating, flooding and air quality. We need to start designing and managing cities as if they are home to an interconnected web of species. Rewilding the city means rethinking all urban space as if they are used and inhabited by humans, plants and animals alike. Everything changes – parking lots, shopping malls, street scapes, parkland, office blocks, residential apartments, suburban gardens, and especially our highways infrastructure. The idea of Half-Earth is gaining ground where 50 per cent of the world's surface is given over to protect biodiversity (Wilson 2016). The same can be applied to the city, although we need to build social and spatial justice into such ideas and ensure they do not promote population control or evictions and displacement of existing communities.

Second, we need to embrace and explore the profound existential challenge to ourselves, unlearning centuries of colonial conquest and the expansion of our industrial world that was built on taming, exploiting and using nature for the sole purpose of a single human species. This means recognizing and enshrining in law, legal personhood for non-human and plant species, whole biomes and bioregions and natural elements, such as forests, mountains and river systems to ensure their regeneration. This also means shifting from a market-based approach to nature management, which can often lead to monetized solutions and displacement, to simply valuing nature for nature's sake (Eisenstein 2018).

Third, finding a new balance with the natural world is not an agenda for scarcity and poverty for humans. It can provide insights into how we restructure our economic and social lives in ways that move away from ceaseless economic growth to a new basis for safety, prosperity and thriving lives. The natural world is a source of abundance, especially if treated like a commons. Key to regaining balance with the natural world is to also regain a social balance amongst the human species. Lessons from social ecologists tell us that if we are able to create more social equality and shared resources, then humans are less likely to exploit nature, and instead protect it (Bookchin 2006). As long as people face poverty and the pressure to compete, nature will be regarded as a free resource that can be used and exploited in any way to improve their lot.

Ideas for Emergency Action

- Legislation can be passed to enshrine in law the rights of non-human entities including landscapes, animals and plants.

- Incentives such as nature prescribing can be used to encourage people to spend more time in nature.

- Schools and colleges can commit to nature-based learning and outdoor learning environments, and workplaces can use time spent restoring nature as part of compulsory staff development.

- Continuous green corridors can be created linking every city neighbourhood, taking land area from highways where needed.

- New wildlife and biodiversity zones can be established on city edges to promote local ecotourism and nature-based learning.

Check out: Half-Earth Project, worldwide; the Forest Schools Association, UK; Rooftop Republic Urban Farming, Hong Kong; Nature Prescriptions, Scotland; Ecuador's "rights of nature" constitutional ruling.

*   *   *

These are our players. The task for all of them is the deep project of learning and unlearning: to unlearn and challenge the instrumental and siloed roles and responsibilities given to us by a society that has brought us to the eleventh hour of our converging emergencies, and to learn new skills, abilities and insights that can respond urgently and decisively to these. This is disruptive and exciting. We need to take from what is already known and workable, and weave this together with what we do not yet know. In the next chapter, I look at the emergency moves our players can make in the urgent quest to save the city.

5

Moves: getting into emergency mode

Without making moves, our energy and drive for change will be wasted. Our moves need to be embedded in our strategic approach and involve the unique strengths, talents and insights of our players. In this chapter I highlight a selection of moves to give a flavour of the bold, necessary and entirely doable changes that can happen right now to save our city. I have focused on the big areas that affect our daily lives where we have to make immediate and decisive changes: how we get around, where we live, what we do for work, how we make decisions. My list is not exhaustive. There will be other emergency moves we need to make and I invite you to develop your own.

In all this we have to make sure our moves are powerful, decisive, immediate and live up to the scale of the triple emergencies, that we see benefits not just in terms of carbon mitigation and adaptation but also social justice and nature recovery. The powerful element is exploring the connections between moves: how change in one area unlocks and accelerates change in another. We have to bring a sense of urgency and purpose to how we think and act. We need immediate changes that also build towards major system change in the decade ahead. The key emergency moves I explore below – car culture, cities economies, placemaking, aviation, nature and democracy – offer opportunities for all our players to act as emergency first responders and use our strategic approach.

Car-ism and how to beat it

Take a look outside your window. There is probably a highway or road – a long stretch of hard surface with painted lines, layered with aggregate and topped with asphalt, a sticky form of crude oil. But what do we actually see and think about when we look at a road? A convenient way to get to work, school and the shopping mall? A corridor of danger if you are a six-year-old or small mammal? Something that gives you a sense of freedom? A challenging obstacle to mobility for the vision impaired that divides you from surrounding communities? A place to meet and flirt, show off your driving skills, or get an adrenaline rush?

The road, the street, the highway, the motorway, the freeway, whatever they are called, are many things to many people. But they have become a naturalized and dominant feature of our cities. They have expanded exponentially, lockstepped with urban growth, seen as an essential building block of how a city grows. The road network plays such a dominant role in city life that the places in between them have almost become incidental. Communities have become the leftover parts, the places you get to using road vehicles, rather than places in themselves. If it cannot be connected by road, it isn't a place.

This is our first, and perhaps most important, emergency move. Central to saving the city and responding to our triple emergencies requires urgently saving it from the car (I use the category of "car" as shorthand to refer to all motor vehicle types). At a very basic level, emissions from road transport, and mainly the private car, are a huge, and still growing, part of the global carbon emissions picture. According to the IPCC (Sims *et al.* 2014) all forms of transport account for about one-quarter of energy-related CO_2 emissions, and transport emissions continue to increase at a faster rate than emissions from other sectors. While we need urgent and aggressive action on transport emissions, it is not cars in themselves that is the focus here. It is a wider mindset that has been built to perpetuate car-based mobility and urbanism – what I call "car-ism".

Like many other "isms", car-ism is a doctrine, a unifying belief and faith that one way of thinking and acting is better than another. It is a dogma that categorises, excludes and oppresses. André Gorz (1973) in his formative essay was clear about the social ideology that the motorcar represented; car-ism is part of an interlocking and complex system that touches on many aspects of how our societies are organized: the geopolitics of oil production, the extractive industries, resource use, exploitative advertising, the increasing role of semi-automated technologies, social status and our sense of self and others, neighbourhood and city planning, debates over employment, debt, finance and corporate control, and the influence of lobbying and market distortion.

Taking part in, or at least accepting, car-ism is a precondition for city life. It is one of the unspoken rules of the game. To reject car-ism is to reject a broader sense of how city life currently operates. Apart from in a few privileged cities in northern Europe, those who reject the car appear as an oddity and an affront to city life. Car-ism is an impressive doctrine when we consider that the urban love affair with the car is only just over a century old. In just over one hundred years, the number of cars has seen spectacular growth – from only a few thousand to over one billion. The early decades of the twentieth century were a transition period, where cars had to compete with other mobility norms built around walking, cycling, mass public transit and horses. In the period after the Second World War, car-ism was pushed to the forefront of the human development model. To develop meant embracing the car. The period from the late 1940s to the 1970s probably saw the greatest consolidation of a single product, and all its associated support industries, that humans have ever achieved. Our urban world was recast through car-ism. In this short time of human history, the motorized vehicle has shaped mobility norms, social status, financial and planning frameworks, leisure and work patterns and broadly accepted cultural and social phenomena such as the rush hour, road rage, ram-raiding and joyriding. Moving around in a car is now an accepted norm. Arrive at a hotel and they will ask you for your vehicle registration plate, go to a meeting and they will tell you where the car parking is.

Tackling car-ism is what I call a keystone activity. Unlocking cities from cars unlocks change in other areas. So much else relies on it, in particular how and where we shop, work and play. Together, these lock in a broader city way of life. It has also brought a basket of negative consequences including road deaths, air pollution, GHG emissions, geopolitical conflict, consumer debt, status anxiety, obesity, social disconnection, the decline of public street life and local economies. They are a major force behind our converging climate, nature and social emergencies. At the very moment when car-ism appears its most dominant we have to rapidly dismantle it.

Reversing out of car-ism means reconnecting with a vision beyond and without cars: creating car-lite and ultimately car-free cities. What city moves will supercharge this? Action in this area goes beyond addressing technical issues of street redesign and highways planning, and even the low hanging fruit of cycle lanes and mass rapid transit systems. These are essential steps and simply need actioning. Given the high levels of carbon emissions tied up in transport, approaching half of the total in some urban areas, cities need a radical rethink of mobility. At the core of this rethink is a focus on eliminating the conditions that make cars necessary. Our aim becomes a socially just, zero-carbon, city-owned mobility planning process that requires implementation across all our city system elements. Some brave policies and financial incentives are coming forward, all of which are doable right now. For example, France is considering a green tax of up to €50,000 (or half the cost of the vehicle) on high-polluting SUVs (Bloomberg 2020), putting disclaimers on car adverts that promote walking and cycling instead, and a €2,500 credit for trading in a car for an e-bike.

One estimate by Friends of the Earth (Hopkinson & Sloman 2018) suggested that car miles might need to be reduced by up to 60 per cent by 2030 to ensure transport emissions play their role in a climate safe future. These reductions by 2030 are a key waymark to ensuring vast swathes of our urban world become car-free by the 2050s. But to get there means year on year reductions in car use in every country and city across the world.

Saving our cities, then, involves comprehensively redesigning out the private car through city-region mass transit linking all settlements, the implementation of pro-walking, car-free neighbourhood planning, micro-mobility options especially e-bikes, creating integrated neighbour-hood transport interchanges, the progressive shrinking and reallocation of road networks as other options come on stream, taking back control of ownership and regulation, and localizing food, work and play. The continued growth of SUVs is a particular problem. Not only do they use more material resources and produce more carbon emissions than regular vehicles, recent research has shown that SUVs are responsible for one-quarter of all pedestrian and cyclist deaths in the United States (Edwards & Leonard 2022). Addressing the growth of SUVs will need con-certed efforts to impose weight and size limits on new passenger vehicles. Tackling car-ism is also about tackling mobility exclusions and inequali-ties, creating an ability to move around that works for everyone regardless of where they live, who they are or how much they earn. The co-benefits of this mobility transition include improved air quality and road safety, bet-ter journey times and reduced costs, stronger local retail, and better health outcomes and quality of life. Let's take a deeper look at two areas where we can start to beat car-ism.

Mass public mobility reimagined

In the main, public finances pay for roads. For example, the European Union has allocated over €80 billion to road building in the last 15 years (European Court of Auditors 2019). This is money accumulated from public taxation. But look again at the road. Who benefits from these public highways? Every day millions of privately built, owned and financed vehicles use them. Transnational vehicle manufacturers – Toyota, Volkswagen, Ford, Daimler, BMW, GM – are all able to accrue multimillion profits every year by selling vehicle units because there are publicly financed roads that they can drive on. Roads, then, are a device for a transfer of public wealth into private hands. Take away this public investment and an infrastructure ceases to exist to facilitate this. Of course, the solution is not to get multinational corporations to pay for their own roads, or to get rid of roads altogether. Rather, humanity needs to wrest back control of roads from just cars and rebalance private and corporate dominated forms of mobility with publicly and collectively owned mass and active forms of transit.

So what would this mean? First, imagine a city where there is a genuine and significant reduction of individual vehicles. This is genuinely difficult to do, as there are so few examples apart from a handful of medieval or canalized northern European cities. But this radical reduction of cars is an essential first step to give back space to other forms of mobility. Life in cities necessitates some level of movement. The issue is how we choose to facilitate it. At the scale of dense populations, mass transit for public use is the most efficient, safe, quickest and cost-effective way to achieve this. But what many of us think about as mass public transport has been so diminished that it is hardly recognisable. It has to be radically reimagined from what it is today.

There are four key elements here. First is integration – it has to be everywhere all the time. Accessible to everyone, regardless of income, gender or race. Second is energy use. Whether heavy or light rail, tram or micro bus, they all have to be fully electric. The supply chains and technologies are all in place. The transition now relies on political will and shifting incentives and subsidies. Third is ownership. Much of the underperformance of our transit system is based on its emergence in an era of deregulation where private operators have been left to design routes and set fares. This privatized system faces many problems including accountability to external shareholders rather than users. But more problematically, in our car-dominated society many bus operators are not able to generate the profit levels needed to expand and face a downward spiral of reduced service, quality and passenger numbers. The solution is municipal-scale employee ownership which offers the scale, integration and accountability to members and users. Local and national state resources need to fund these activities that help mount emergency-level responses. In terms of mobility, this will involve a transfer of subsidies and public funds from private to public options, making mass transit more financially viable.

The final aspect is affordability. This is the most challenging as good quality mass public mobility comes at a significant cost. But we need to regard our ability to move around as a universal basic service just like healthcare, energy provision and housing. Quite simply, moving people around needs to become a much bigger slice of government spending, and where that government spending does not exist, especially in lower-income countries, then international banks need to subsidize and build infrastructure. Tonnes of carbon taken out of a city transport system, wherever it is in the world, means a safer climate for everyone in the long run. Free public transport initiatives are starting to grow across the world. The key lesson is that making something free does not work on its own to get people out of their cars. It tends to benefit those who already use public transit. However, offering it for free is the incentive that can justify restraining car use through, for example, use-charging, no-drive zones and parking levies. Once there is a free, fully integrated electric mobility alternative, city politicians everywhere can start to regulate and ultimately ban car use at scale. In parallel, additional investments will be needed to make them safe and reliable, especially in terms of increasing public confidence, post-pandemic, of being in close proximity to others, as well as making people comfortable to travel on their own and at night.

The 15-minute neighbourhood

The second big area of action is simply to move less and thereby live more fulfilling local lives. This has become popularized recently through the idea of the 15-minute neighbourhood – a simple proposition that states that one's daily needs can be attained within a 15-minute walk, or even shorter cycle ride, from one's home (Emery & Thrift 2021). Fifteen minutes is not a magic time limit; 20 minutes is also used, and 30 may also be reasonable. It is important to note that the 15-minute neighbourhood idea does not aim to restrict people's freedom of movement and keep them in their immediate neighbourhoods. The point is simply to replace local short journeys that are currently made by car with walking or cycling, while still acknowledging people will still move longer distances for other reasons and by other means. While it does impose some restrictions on very local car circulation, these are outweighed by new freedoms offered to other groups especially families, the elderly, children and those with complex mobility needs.

Central to facilitating such shifts are increased building densities and an active travel network: wider pavements, safe crossing points, road filtering and closures, and cycle lanes across every street in a locality. Through such changes, walking and cycling become the quickest, safest and cheapest options, with motor vehicles displaced around the edges. Innovators and policy makers often talk about micro-mobility: the myriad ways to get about locally propelled by foot, cycles, e-bikes, scooters, unicycles, skateboards, or mobility chairs. The key aspect to micro-mobility is that these are small scale, appropriate to the local level and can be brought forward much more quickly and at lower costs than large mass transit projects, which require significant investment of time and money.

But the 15-minute neighbourhood will not work in isolation. We need to avoid creating disconnected gentrified bubbles, where 15-minute neighbourhoods emerge around existing hotspots of local services in higher-income neighbourhoods. The 15-minute neighbourhood needs to be part of a larger plan for a 15-minute city – an overall denser city, with lower car-use, and where traffic-free neighbourhoods make sense internally but are also well connected. Unlocked from the car and its road network, cities can be a mosaic of interconnected neighbourhoods. There will still be a need for some vehicle use for those with accessibility issues and larger deliveries. But the key aspect is to change the underpinning neighbourhood economy by creating localized employment hubs and affordable consumer options to reduce commuting and distribution. Larger e-bikes, cargo bikes, trailers and e-box bikes can take up a significant proportion of the flow of items around a neighbourhood.

Our discussions about car-ism therefore takes us quickly into exploring the city economy. We have to take a forensic look at our movement patterns. Why are we moving around in the first place? Who is travelling further than they need? Are employment hubs built near residential areas or on motorway intersections and mega distribution centres? Who is locked in to precarious and low-paid work far from their home? What forms of mobility make us feel safe and less safe? What flow of goods and services rely on very long supply chains beyond localities?

Car-ism, then, has created the opposite conditions of the 15-minute city – the one-hour city. It has allowed residents to seek out more and more niche goods, services and activities across larger distances. The car and its ever-expanding road network has been one of the central conditions for urban society to become less equal and more socially different. Social groups can use their unequal access to car-based mobility to seek out specialist work, leisure and retail options that are more tailored to who they are or want to be, what goods and services they desire, rather than relying on what is available locally, and what kinds of people, schools and facilities they want to associate with. In turn this has created localities that are temporary transit points in a broader urban society based around the car.

Ideas for emergency action

- Create a strategic plan to make your city car free in the next ten years. Spend time engaging the public on the plan to ensure buy in.

- Adopt Vision Zero Targets for road deaths and enforce blanket lower speed limits across all urban roads.

- Bring mass transit back into public ownership, make it free and fully electric; commit to routes and timetables that are continuous and expansive.

- Shrink the city highway system by 5 per cent a year, with the aim of halving its size in ten years. Use released land for biodiversity, food, active travel and carbon storage.

- Replan your city as a network of 15-minute neighbourhoods based on dense interconnected active travel networks.

- Use incentives such as workplace parking levies, congestion charging, use charges, ultra-low emission zones and local municipal bonds to raise money to pay for changes.

- Create car exclusion zones around all public buildings and schools.

- Get rid of zoning that separate work, retail, and residential and bring them all together.

- Invest in micro-mobility hubs, bike and car sharing across the city.

Check out: free buses in Tallin, Dunkirk and Luxemburg; Ghent's car free city plan; Barcelona's superblocks; Dia sin Carro y Moto (Motor and car free day), Bogota; Malmos' Cykelköket (bike kitchen); Milan's Cambio Cycling mobility plan.

The new city economy

One aspect that unites almost all city residents, whether recently arrived or long-term inhabitant, is a desire for a decent livelihood. We can distil this into a very straightforward challenge: how can we all live well in cities within the natural limits of our planet and without compromising the ability of others to live well? I use the term livelihood, rather than simply work, for a reason. Livelihood refers to a range of aspects we need in our lives to live well (Rakodi & Lloyd-Jones 2002). Formal work is certainly part of this, but it is not the only part. Much of what underpins a good livelihood is down to what researchers and policy makers call "capabilities". The philosopher Martha Nussbaum (2000), for example, has stressed there are a range of core capabilities that ensure a good life and should be the focus for our policy and action. These include key issues such as work, health, recognition and the freedom of expression, but also political and economic control over our lives. These capabilities offer us the potential not just of livelihoods, but of thriving livelihoods. In our quest to save the city, we are creating the conditions for thriving cities, where everyone can develop and meet their potential without harming others or the planet we all depend upon.

We can look at the city through this thriving livelihoods perspective. Our current urban economic model undoubtedly brings benefits: abundant consumer goods and services, employment and education opportunities, creativity and innovation. But many of these benefits are fleeting, tied up with social status, and are deeply geographically uneven.

Decades of competition, privatization, precarious labour and free market economics has created a basket of winners and losers, pitting places against each other in a race to the bottom. Global investors have a free hand to buy up land and assets, sucking up and extracting community wealth to distant boardrooms. Most worrying, there are now serious risks emerging at a planetary level through global heating and biodiversity loss.

Some major changes to the way city economies work, then, are long overdue. But the problem we face is imagining how it could be different. The economy is closely connected to our sense of ability to live well. Interfering with it feels risky and dangerous. Even talking about changing the economy raises concerns of authoritarian communism, permanent recession, or mass unemployment. It usually feels safer to leave it alone in all its imperfection. Rather than a major overhaul, the answer is often seen as more of the same; that given more time or better management, the current economic model will work to secure good livelihoods. This has been popularized in neoliberal economic ideology through the idea of "trickle down": that the function of the city economy is to facilitate conditions for wealth creators and inward investors through tax breaks and other incentives that ultimately will create prosperity which trickles down and benefits lower-income groups and communities. The idea that "a rising tide lifts all boats" is strongly ingrained in everyday discourse and policy. It is easy to see why this is popular, as people are keen to have a share of the wealth they see around them. In fact, hardly anything like this happens (Raworth 2017). In reality, wealth is concentrated upwards and outwards through external procurement and shareholder dividends, leaving few benefits locally to those who actually supported that wealth generation in the first place. The contemporary city is left with a growing gap between the haves and the have nots, a deep and lasting sense that for many urban residents, city development is not benefiting them.

Our task is to challenge this taken for granted story. We need to refocus the job of a city economy not at the service of perpetual economic growth, but at creating thriving livelihoods for all. An economy that is valued for how much it just grows does not make sense on many levels. A constantly growing economy uses more raw materials than the planet can sustain, accumulates carbon in the atmosphere that is driving dangerous heating, and is creating unprecedented levels of planetary inequality. The reality of an economy that continues to grow also does not add up. Society tends to value increases in GDP as a good measure of the health of the economy and increases of 3 per cent per year are typically required to ensure there is money to reinvest. This would see the economy becoming several times bigger by the end of this century. Given, we are already overshooting planetary boundaries with the current size of the economy, one that is several times larger would mean levels of global heating, resource use and biodiversity loss that would largely end life for humans (Hickel 2021). What would a new city economy look like that ensures thriving lives for all within planetary boundaries? Below I take a look at two features.

City wealth building

One of the hallmarks of our current economy is the speed and volatility of money and resources that can easily flow in and out of cities, facilitated by growing privatization that has shaped the way we run our economies and provide goods and services over the last 40 years. If you don't own it, you can't control it. Land deals, procurement contracts, employment conditions are increasingly controlled in distant boardrooms legally answerable to shareholders seeking to maximize their return on investment. While these external connections do bring investment, ideas and skills, they also extract and drain resources. Imagine a city economy as a bucket, but one that is full of holes. As money comes in at the top, it simply trickles out of the bottom. So the task is to plug the leaks and build a deep reservoir of community wealth that can be used to build resilience and thriving livelihoods (Ward & Lewis 2002). We need to find a new balance so communities can control and benefit from the immense resources that flow across our world every day.

Community wealth building is an idea that has been gaining momentum for some years now (Lloyd Goodwin et al. 2021). Since the pioneering worker-owned cooperatives in Cleveland, Ohio in the late 1960s, activists and policy makers have been exploring how to make money stick and circulate within local economies rather than being shaped by corporate priorities. Community wealth building is based around identifying anchor institutions like hospitals, schools and colleges that can trade goods and services locally. This can kickstart employment and income generation in a place, and brings further benefits if it is based around community-owned and managed businesses in lower-income areas traditionally cut off from wealth making in cities. For example, community-minded businesses also seek to provide jobs that are better paid with good conditions, protect the local environment, reduce energy use and carbon emissions.

One way of exploring what this means in practice comes from the idea of the foundational or substantive economy, that part of the economy that supports and underpins our well-being and ability to live well (Bärnthaler, Novy & Plank 2021). There are certain areas that create a foundation for our ability to live well, such as decent and affordable housing, an ability to move around, or access sufficient energy. They need to be under the influence and ownership of bodies accountable to the public and the communities they serve, and operate within the limits of our natural world. Reframed like this, the city economy becomes a source of thriving and dignified livelihoods for all as well as an immense site of civic innovation and bulwark against the global casino economy.

Civil society is bursting with potential, initiatives and skills which can build community wealth and mount effective responses to the triple emergencies. Features of this new thriving economy include collaborative peer2peer production, open-source data, flat governance structures, mutualism, co-operative and employee-owned firms, anchor organizations and circular, regenerative activities (Gruszka 2017; Fath *et al.* 2019). Examples in practice include citizen housing, which can be built or retrofitted using local materials and labour, community ownership of assets such as grocery stores or pubs, local procurement of basic goods and services like cleaning and laundry services by anchor institutions, community-based and open-source digital manufacture that can stimulate local construction and employment, neighbourhood enterprises and maker spaces. Co-operative and employee-owned enterprises are a central pillar in this community wealth building. They are able to lock in workplace democracy, instinctively respond to community needs and pay living wages to employees while avoiding the huge disparities of pay seen in the corporate world. Community initiatives can come together to create wider momentum at the city level. New city-wide institutions, for example, can retrain and reskill construction industry workers to roll out house by house retrofit insulation programmes; novel municipal finance initiatives can provide affordable lending to stimulate community-based investment; and, co-operatives can trade amongst each other creating more opportunities for investment and employment.

The future of work

Work is important to our lives. It provides a source of income and generates a sense of worth, direction and achievement. But the world of work is deeply divided and poorly set up to tackle the challenges ahead. It can be a source of social inequality, mundanity, precarity, harassment and discrimination, and it can drive excess consumption of goods and services as well as anxiety and depression. And for a small minority – the rentier class – work is largely unproductive and simply equates to charging rents and extracting surplus from others simply by owning assets and resources (Christophers 2022). Increasingly, what actually drives the economy is the globalization of financial capitalism rather than activity in the real productive economy (Guoping & Zhou 2015).

Our task in building a new city economy is to take back control of work, who creates it, and what impact it has on us and the planet. We have to start by doing away with what harms. Debates around the meaning of work are not new. In 1884, visionary socialist William Morris wrote about the distinction between useful work and useless toil as he saw the drudgery involved in the industrial cities that were growing around him. More recently, John Holloway (2010) has made the distinction between alienated "labour", where we sell our labour time for the benefit of others, and "doing", where we are free to dispose of our time in the way we choose. The late radical sociologist David Graeber (2019) pointed to the rise of "bullshit jobs" that have ceased to offer much in the way of social value across a range of sectors, including advertising, armaments manufacture or investment finance. Creating a sense of what constitutes good work, then, is important.

But there are huge changes unfolding in the world of work. Globally, as more and more people are brought into the formal economy, there may be less work, especially in more established economies. Technological advances in automation and AI are shedding labour at a faster pace than new jobs are created. Lower paid, less stable jobs in the gig economy are increasing, with precarious workers servicing a more stable salaried class. How we all find meaningful livelihoods amongst this rapid change and respond to our converging emergencies is a defining challenge of our age.

But a groundswell of new approaches is being developed to rethink work in the new city economy. The idea of a four-day week is gaining momentum as a way to distribute work more evenly between those who have not enough and those who have too much (Stronge & Harper 2019). Sharing work more evenly has positive impacts in terms of self-worth, fulfilment and work–life balance. Redistribution of work is a key pillar in our quest for greater social justice, and also in reducing excess consumption of those with the highest salaries. But this redistribution needs to go hand in hand with broader changes to how we remunerate work. The surest way to create social stability is to guarantee people a certain level of income. A "universal basic income" (UBI) – an unconditional, automatic non-means tested payment to every individual as a right of citizenship – is an idea that is gaining considerable attention and is changing people's relationship to work (Lansley & Reed 2019). Many cities across the world are undertaking basic income trials. Providing everyone of work age a guaranteed minimum income provides a universal and unconditional safety net that can offer dignity and security, but crucially it also allows people to explore more sustainable living as they work less.

While we need to be freed from low-paid and unfulfilling work, we also need to be freed from debt and the increasing cost of living. Shifting large parts of the foundational city economy, like food, energy and housing provision, into public and community ownership can help control prices and reduce everyday costs. But we also need a city "debt jubilee": a mass cancellation of debt, especially for low-income groups, who find themselves in a negative spiral of debt repayments. Freeing people from debt, especially associated with housing, creates the possibility of freedom from precarious and low-paid work, and gives them back choice and control, which can include the option to live slower, more local and sustainable lives.

Finally, this all comes together by creating a "Green New Deal" at the city level. The idea references Roosevelt's New Deal of the 1930s which put the United States back to work and got the country out of recession. But now, this is a New Deal for our converging emergencies. It is an idea that has been taken in exciting new directions through student and youth groups such as New Deal Rising, as well as in the congressional resolution led by Alexandria Ocasio-Cortez, which offers jobs, justice and decarbonization (Ocasio-Cortez 2019). These plans are based on harnessing the formidable powers and resources of the state in tackling the challenges ahead and directing them to community needs. Key to this will be city-level deals where vast sums are allocated to green transport, housing, food and energy, all of which create a new jobs bonanza. The resources required are secured through progressive taxation, closing down on tax avoidance, shifting resources, for example away from militarism, as well as borrowing from the international money markets through green and community-focused quantitative easing (Pettifor 2019). At its heart it is a plan to transform our economies and city work by creating high-paid, green jobs. Job to job transitions will be required to meet the task of zero-carbon, socially-just, nature-friendly work. Trade union partners will play a key role here supporting and promoting the emergence of new work patterns.

Ideas for emergency action

- New approaches to measuring city progress can be adopted that do not just focus on economic growth, but instead focus on rapid and large-scale carbon reduction, nature restoration, poverty reduction and social equality.

- Cities can formally adopt community wealth-building as a centre pierce for prosperity and well-being.

- Four-day week trials can address work–life balance, with knock-on effects of increased well-being, reduced consumerism, commuting and energy use.

- Universal basic income and income guarantees can be used to ensure people do not fall below a minimum threshold.

- Key city services such as water, food, transport and housing can be reconceived as universal basic services offering universally free access to aspects that underpin our well-being.

- Cities can adopt a Green New Deal to stimulate investment and jobs in green transition areas.

- Unions and other civic movements need to play a central role with city authorities to ensure a just transition and equitable job-to-job transitions for workers.

Check out: 4-Day Week Trial, UK; Citizen Income pilots,
Utrecht, Barcelona, Oakland; Green New Deal programmes,
London, Glasgow, Seattle; Rio de Janeiro's Social Progress Index;
Everyone Every Day community enterprise hub, Barking
and Dagenham.

Emergency placemaking

The places we live are a source of our daily well-being. They are the backdrop to our lives, our families and friends, where we return to, dwell and thrive. Making places is an ancient art. It takes time and patience for the texture, depth and meaning of a place to emerge. It takes effort, interventions, good regulation and planning and meaningful dialogue. But over the last few decades, places have been shaped more by the logic of financial profit and the needs of our industrial and fossil fuel-based society than by people, nature and community. This means particular things on the ground. First, it creates increasing precarity and division between the haves and the have nots. Gentrification further drives these divisions with lower-income groups pushed out of traditional neighbourhoods, and new-build apartments built in areas cleared of former industrial uses (Gould & Lewis 2012). This is now a process seen in cities across the world and fuels the growth of speculative investments in property markets that seek quick gains from selling housing at the highest price. Those on lower incomes continue to be displaced from higher-cost areas and pushed down and out of the urban hierarchy. Global private equity firms then have a free hand to buy up urban land and maximize yields for shareholders and investors (Lees, Shin & López-Morales 2016).

A second consequence has been the increasing informality to housing as millions of people across the world live in informal settlements, what are popularly and prejudicially known as slums (Gouverneur 2015). Many emerge chaotically and illegally, yet over time become more formalized, although without guaranteeing their inhabitants access to clean water, transport, energy and work. Finally, this precarity and informality encourages separation and segregation: gated communities, suburban extensions and private luxury condominiums. Our urban places are increasingly geographically sorted and divided by income, age and ethnicity. Localities are becoming more internally homogenous, wary of each other and the differences they pose. This has implications for how we see ourselves, others, and the places we live, and how we view and accept displaced and migrating people.

Given these tendencies, as we face the converging nature, social and climate emergencies, we have to take a fresh, urgent and profound look at how we make places. I want to highlight three placemaking ingredients that will help us navigate the triple emergencies. The first is resilience. While resilience has a technical meaning associated with withstanding shocks, it goes much deeper than this; it is the ability to build dense social networks that protect, nurture and support (Chandler & Reid 2016). On one level, place resilience is about dealing with what gets thrown at us. Building in redundancies is important: if one part of the system fails then other parts can stand in and the whole system does not collapse. But resilience is also about place politics: organizing to change the conditions of the (neoliberal) city system to make sure shocks and instability are reduced in the first place, or at least reducing unequal impacts across social groups (MacKinnon & Derickson 2012).

The second is safety. This operates on a number of levels. It means safety from harm in its many forms, such as pollution, road accidents, dangerous housing, workplace risks, discrimination and harassment, forced labour and modern slavery. People's varied experiences and characteristics vary dramatically, and these need to be understood and factored in to how we make places. Avoiding amplifying or reinforcing existing vulnerabilities is a really important lesson for placemakers. One of the most significant areas is that of climate safety. A whole raft of unsafe conditions is unfolding due to climate breakdown. Each place needs to play their part in the great mitigation challenge of keeping the increase in global temperatures to below 1.5°C, as well as adapting to an Earth system already showing signs of breakdown. But since many of the unsafe conditions of dangerous heating are already with us, or locked into current global heating projections, we need to focus on adaptions to make places safer, whether protection from extreme heat, flooding, or drought.

The third key ingredient is a collection of intentions around justice, equality and equity. They are all subtly different, but complementary. Justice is a broad set of demands which can cover recognition, rights or access to political and legal processes. Equality is about having an equal and fair shares approach to people, resources and processes; while equity allocates on the basis of need (Minow 2021). Understanding these issues is key for placemakers. Some people simply need more to make up for structural issues that might exclude, marginalize or oppress. For example, the needs of an eight-year-old child or blind retired person require prioritizing over those of a physically able working-age adult. Let's look at a couple of activities – zero-carbon places and land reclamation – that can be implemented right now to shift placemaking into emergency mode, and to build places that are resilient, safe and equitable.

Zero-carbon places

Current placemaking strategies lock urban areas into global heating through the use of high carbon materials like steel and concrete, car-based dependency, poor street and building design, inefficient energy use, sprawling development and a separation between work, living, food and leisure. High emission places also usually mean other problems are close at hand such as lack of democratic control, low levels of social interaction and reduced greenspace and social divisions.

Zero-emissions emergency placemaking is an exciting new paradigm for city building with social equality, nature-based solutions, and car-free design at its core. It will require the skills and insights of all our players and touches on all our strategy areas. Each place needs to take its responsibility seriously and not pass the burden of carbon reduction on to others elsewhere or in the future. Given the immensity of the carbon reduction challenge, negative carbon strategies will also be required. This will mean large-scale reallocation of land to activities that can draw down carbon from the atmosphere (Hawken 2018). Fundamental to zero-carbon place-making is ensuring that buildings and infrastructures emit as little carbon as possible across their lifetime, including operational heating and cooling needs, the materials they are built from at construction phase, and any carbon implications from repair and deconstruction. Rigorous technical building standards and codes are essential to reduce the energy needs and carbon impact of housing. Exciting examples are emerging such as the Living Building Challenge standard (Forsberg & de Souza 2021) which uses seven holistic design challenges to create restorative and regenerative places. Reliance on concrete and steel construction is a central challenge to be addressed. According to researchers at Imperial College London, cement production and iron/steel manufacture each create over two billion tonnes of carbon dioxide per year – around 14 per cent of global CO_2 emissions (Fennell *et al.* 2022). Natural materials such as wood, hemp and lime naturally sequester GHGs and provide workable alternatives to these building-related emissions.

Beyond energy efficiency gains and lower carbon technologies, the wider challenge is to embark on a civic energy revolution based on municipally-owned, renewable and affordable energy supply. All forms of generated energy are a precious investment of time, effort and resources, so our first priority is to reduce demand in order to simply use less energy. For the remainder that we do need, municipal energy companies drawing on pioneering examples such as Germany's Stadtwerke show how to use city resources to harvest as much renewable energy as possible. At the same time innovations need rapidly scaling, including distributed energy networks, district heating, local smart grids, community energy, zero emissions community-led developments, Combined Heat and Power (CHP), onshore wind, solar photovoltaics, anaerobic digestion, battery storage, and the new skills that will underpin these (Cowtan 2017). We need to start to think about homes and neighbourhoods as mini power stations, generating renewable energy and feeding in to local energy networks to build resilience and affordability (Rifkin 2013). Community and city ownership means that energy costs are more affordable and dividends locally reinvested.

The biggest challenge to address is the way we live in places and the way they function. Zero-carbon placemaking regards places as complex and integrated wholes. Architect Jan Gehl (2008) urged us to focus on the life between buildings. We must begin to forensically take apart life in a place and reconstruct it in a way fitting for the emergencies ahead. For example, just as important as building energy efficiency is what we are actually doing in buildings, how we relate to and treat each other, and how and why we move around our places. There is little point in having A-Rated buildings, if they facilitate F-Rated lifestyles. How and why do we construct daily itineraries? Are there well-paid and meaningful jobs available locally, and employment hubs where they can be located? Are there district centres with local services, and are these controlled locally so independent businesses can thrive and compete with multinationals? Is there an integrated active travel network so people can walk and cycle to local services? Is there a local transport interchange which allows movement between places? Is enough green space dedicated to food growing, biodiversity and leisure? These are pressing issues for those who set the conditions and resources for how places are made.

Clearly, there are many changes already unfolding. The shift to home-working especially during the Covid-19 pandemic has given an extra boost to many places, and has highlighted the importance of local services and green spaces. But it has also highlighted the inadequate and cramped living conditions for many and the increasing dependence on smart and semi-automated technologies in the home and in the community, all embedded in an energy bloated global infrastructure of data storage, media streaming and online meeting platforms. As James Bridle (2018) points out, the energy and cooling requirements of planetary data systems will shortly have a carbon footprint equal to the airline industry.

Reclaiming land

One of the longstanding barriers to saving the city is land ownership. Land is a finite commodity. There is only so much of it. So it has to be put to really good use and managed effectively. The history of the last few hundred years has been one of enclosure and the increase of private ownership (Hayes 2021). This continues to the present day. Guy Shrubsole (2020) found two-thirds of land in the UK is owned by less than 0.5 per cent of the population, while over 20 million families share its urban land. The city is a product of this deeply unequal allocation of land and changing this will be crucial to saving the city.

As something becomes privately owned, it is enclosed in the sense that it is not available for use by others. That owner is now able to control and restrict access, charge for its use, offer access to the highest bidder, and determine it use. This process of privatization and enclosure has been central to how our capitalist economy has evolved over time (Linebaugh 2014). When industrialists run out of resources to own and use, their rate of profit falls. They move on to new places to access lower-cost workers and resources to maintain profit levels. This is what is happening in our cities today. More and more public and community land is privatized, bought up by overseas and offshore private equity firms or hedge and pension funds, then sold and resold as if it were a commodity like oil or wheat. Placemakers and planners need to find ways to stop this cycle of land speculation and put land back into public and community ownership.

This makes sense on many levels. Communities and public bodies are answerable to a larger group of people compared to private entities. This gives greater accountability and scrutiny over how land is used. Community control also leads to uses that are more beneficial for the community, through activities that are more affordable, equal, resilient and safe. With greater access to land and an ability to manage how it is used, communities can also generate activities that can directly tackle the triple emergencies, including local food growing and renewable energy, community transport, play and leisure spaces, and places for independent trade and employment. Once land is locally controlled, a new approach to house building and neighbourhood creation can emerge. In work with housing researchers (Lang, Chatterton & Mullins 2020), we found that community builders are pioneering new forms of people-powered home making and funding through citizen shares, open-source design, collaborative manufacture and community ownership which also bring a range of co-benefits around community energy, food and transport.

Significantly, there will need to be a large-scale reallocation of land to support greenhouse gas removal, nature rewilding and recovery and the broader shift to plant-based diets. Planning frameworks will need to be redesigned around metrics linked to city carbon budgets and the United Nation's Sustainable Development Goals. Every single planning decision will need to be assessed against emergency metrics: its impact on the city carbon budget, social justice and nature recovery. Placing a carbon value on all new developments, with financial penalties for excess use and costs imposed for future damage, will also steer projects towards zero-carbon emissions. Drawing on windfalls from changes in taxation and subsidies, especially around taxing land and not income, municipalities can buy and transfer land into public and community organizations (Monbiot *et al.* 2019).

The preservation and extension of green space is central to rethinking land. Building in biodiversity gains, blue-green infrastructure, biodiversity corridors and rewilding are part of tackling the nature emergency. This will mean increasing the density of residential areas in cities, reconfiguring suburbs and exurban sprawl, decommissioning out of town and strip malls as local district centres come back into use, opening up waterways, closing highways, and creating under and over passes so non-human species can once again roam free.

Ideas for emergency action

- Local businesses and colleges can create a city-wide housing retrofit programme, backed by loans from municipal bonds.

- City-owned energy companies can commit to 100 per cent renewable energy production sold at affordable levels.

- Derelict under-used land can be placed in community-owned land trusts.

- City ordinances can reduce and eliminate land banking and hoarding by speculative developers.

- City authorities can create regulations to prioritize and expand community-led housing developments.

- Strict design codes can be used to ensure all aspects of place making meet our triple emergencies in terms of carbon, biodiversity and affordability.

Check out: Stadtwerk Munchen energy company; Repowering London community energy; Te Kura Whare Living Building community centre, New Zealand; More than Housing Cooperative Zurich; Hamburg's Gängeviertel quarter; Champlain Housing Trust, Vermont; Abahlali baseMjondolo shack dwellers' movement, South Africa.

Your flight is cancelled

Aviation is one of the spectacular developments of the recent human age. Through access to gigantic fossil fuel energy reserves, we have created an ability to travel around the world at heights, distances and speeds unknown in human history. But the unfortunate truth is that this amazing achievement comes at a high cost for people and planet. While aviation is only a small part of the global carbon emissions picture, it is a fast-growing sector without a viable and safe plan to respond to the triple emergencies. As the world decarbonizes, aviation will be the outlier, stubbornly expanding, burning carbon without a strategy for switching to renewable sources. If by the 2030s other city sectors have halved their greenhouse gas emissions, aviation will be the main source of global heating in cities that have airports. This is a huge concern. The aviation industry is increasingly using a novel form of greenwashing – or jetwashing – to create public confidence in technologies such as sustainable fuels and electric flights as viable strategies compatible with tackling climate breakdown (Peeters *et al.* 2016). While only a small fraction of the world's population has access to this form of transport, battles over aviation will become one of the pivotal moments of the mid twenty-first century. So is a jet-zero future possible? Let's take a look at what is going on and what the challenge is.

In the post-Second World War era, the international airline industry has grown at a phenomenal rate. At the start of jet travel in the 1960s, around 100 million people travelled by air each year. In 2019 global passenger numbers had peaked at 4.5 billion per year (ICAO 2019). While this figure almost halved by 2022 as a result of the Covid-19 pandemic, the aviation industry plans for a return to pre-pandemic numbers. In spite of the Covid-19 downturn, a significant part of our urban world has become connected to a truly global aviation network. Much of this growth has been in the Asia-Pacific region, with China and India seeing a huge expansion in domestic flights. To get a total picture of the extent of growth we need to also consider the parallel rise of the air freight industry which now carries around 50 million tonnes of goods per year, and the renewed growth, especially post-pandemic, of smaller, private luxury air travel, with 500 operators now providing private charter services (Gollan 2022).

This is also a story of the love affair between urbanization and aviation. Urban growth has become lock-stepped with the aviation industry and the desire for global connectivity. Airports are classic urban mega-projects, creating a whole raft of associated infrastructure including hotels, roads, rail, employment zones and leisure facilities. They are key growth poles in the broader pro-growth city paradigm. While global city airports continue to expand, the growth of regional airports that serve small and medium cities has been even more dramatic. These have facilitated the emergence of "city breaks", the novel and deeply damaging idea that you can pop to a city thousands of miles away for a weekend vacation. Rapidly growing airline companies and airport operators have developed and supported the emergence of airports in less popular locations to open the reach for air travel further. Eager not to be left behind, cities compete with each other to attract them. The highpoint of the fossil fuel age for cities, then, is to attract an airport and its airline operators.

The rocky road to jet zero

The global airline industry presents us with a number of key challenges that inform our emergency moves to save the city. The first concerns geography and our relationship to it: the continued compression of space for a tiny, in global terms, travelling elite, the erosion of meaningful geography between these spaces, and the rapid and superficial consumption of gentrified and commodified city spaces by air travellers. City space is being reconfigured to service this air travelling elite through associated infrastructure of highways, corporate hotels, boutique shopping, luxury consumption, airport employment zones and eateries. Hardly any of this is connected to or rooted in local community economies, or serves green jobs. The uncomfortable fact for a small global minority who have enjoyed international tourism for decades is that, as currently conceived, it can never be truly sustainable in a way that responds to our converging emergencies (Hall, Scott & Gössling 2013).

Second is the level of fossil fuel use, in this case mainly kerosene, and the associated carbon emissions. Moving a mass of people, things and planes around the world is incredibly fuel dense. Just before the Covid-19 pandemic, the global airline industry hit an all-time high in terms of fossil fuel use at 95 billion gallons of jet fuel (Burgueño Salas 2022). Burning this level of fossil fuel creates about one billion tonnes of carbon emissions, which in turn has a direct impact on global heating. These emissions account for slightly more than 2 per cent of total anthropogenic global heating (Ritchie 2020b). This is smaller than road transport which accounts for about 15 per cent of total global carbon emissions (Ritchie 2020a). But there are qualifiers to this impact. Two per cent of the global total is still a huge figure and bigger than the total carbon emissions of many countries. Further, the global heating impact of global aviation is underestimated. Airplanes leave behind condensation trails – high altitude water mixed with soot – which have a non-CO_2 effect in terms of global heating that could actually double the heating impact of aviation (Kärcher 2018). This full impact on climate heating is not currently accounted for by industry or government. Researchers have found that the impact of contrails could increase three-fold over the period to 2050, which will outweigh any efforts to limit the impact of aviation on global heating (Bock & Burkhardt 2019). Reducing these impacts are possible by changing flight routes or heights, but none of these are currently being implemented.

These are significant issues, especially for a sector that does not have a viable plan for reducing or dealing with its emissions. Its overall intention is to reduce carbon emissions by half by 2050, an ambition that is not currently technically feasible, and most worryingly is not even in line with the Paris Agreement targets. This is in contrast to road transport where active planning is underway to decarbonize vehicles through mass electrification, as well as curtailing and replacing vehicle use with active travel and mass transit.

Third, it is not just the impact of the actual flight. We also have to factor in the associated development – how the growth of aviation supports further urban growth. When an airport is built and passengers or freight arrive it creates concrete, asphalt and steel construction and retail outlets. All of this continues to lock our urban areas into a high-growth, consumer-oriented model that is energy dependent, carbon emitting and drives global temperature rises. Returning to our strategic approach, we have to think systematically and consider all the associated effects of the "aviation system". There are also a range of other issues from aviation that affect cities, especially noise pollution, which is deeply disruptive to humans and non-humans alike.

Fourth, there are significant social inequalities built into flying. Unlike road transport which, although riven with social inequalities, is relatively accessible globally, aviation serves an even smaller and higher-income section of the population, especially when put into a global context. The entirety of Africa, for example, only accounted for around 2 per cent of all airline passengers in 2018 and ten countries accounted for about 60 per cent of total aviation CO_2 (Possible 2021). Although 10 million people fly every day, it is still a minority pursuit. Moreover, it is frequent fliers that are having the greatest impact, those small numbers of passengers who are making multiple trips. Data shows that in the UK 15 per cent of the population make up 70 per cent of passenger numbers, and this is a pattern replicated across most developed economies (Hopkinson & Cairns 2021).

There are also key concerns about the number of jobs aviation provides. Some estimates suggest that, globally, around 11 million people are employed directly in the sector. This is a significant amount. Our strategy would celebrate this and explore direct job to job transitions for these essential infrastructure workers and seek to utilize their existing knowledge to support the emergence of sustainable travel options. But the airport sector also overstates the argument on jobs. Jobs generated in this sector come at a high price, often subsidized by tax receipts and other publicly financed incentives to create employment hubs near airports. But many are lower skilled warehouse jobs rather than the hoped-for high-tech aviation jobs. In addition, aviation employment zones are usually car dependent and not well served by mass transit, locking in workers to high carbon-emitting long vehicle journeys. Public subsidies would be better spent investing in community-based employment hubs in sectors that also address biodiversity loss, green energy or carbon capture. Moreover, jobs related to aviation, especially in the global tourist or freight industries, lead to further emissions from the consumerism and urban sprawl associated with resorts.

Finally, there is the issue of technological change. Since the age of jet travel in the 1960s, there have been huge efficiency gains related to advances in aircraft weight, materials, flight techniques and fuel use. These will undoubtedly continue. But the issue we face is that these efficiency gains cannot happen fast enough and are eaten up by the continued growth in the size and number of airports and passenger numbers. As more nodes become connected to the global aviation web, there are more planes, more passengers, more air freight, more mass tourism, more long-distance city breaks, more consumerism.

Much of the reason why global aviation is committed to continued growth is that it is bullish about the prospect of what it calls sustainable aviation fuels. There is no doubt that non-fossil fuel alternatives to kerosene, including waste fuels and biofuels, will play some role in future aviation. The industry is planning for sustainable biofuels to meet around 10 per cent of aviation fuel needs by 2030, rising to 50 per cent by 2040 (World Economic Forum 2021). But even this is not fast enough given we have to eradicate all fossil fuel use as soon as possible. While alternative biofuels are carbon neutral at face value, given that the aviation industry needs billions of gallons of fuel every year, it is difficult to imagine allocating farm land to meet this demand. This is especially acute if we consider the pressure to ringfence land for rewilding, biodiversity corridors, food production and carbon sequestration, as well as competition for biofuels from other transport (road) and home heating, and the likely supply disruptions from war and climate-induced instabilities. Biofuel production also accelerates existing problems around the spread of monoculture plantations, deforestation and the displacement of Indigenous peoples (Friends of the Earth 2021).

Plans for electrification are even more speculative. Smaller electric planes already exist. Finland, for example, is trialling small light electric aircraft on short-haul routes. The question is what scale of travel and size of aircraft can electric air travel support, and in what timescale? It is not likely that technological advances can provide battery energy storage any time soon at the size and scale of mass aviation that people have become accustomed to (Bauen *et al.* 2020). Nor can full electric propulsion come on stream fast enough to curb the growth of fossil fuel aviation emissions.

Building momentum against aviation

What moves do we need to take in this context? The human desire for travel will continue. But in our decade of transformation, the task is to bring air travel back into the realms of safety. So what should particular cities do? Aviation, like many sectors that burn lots of carbon, presents us with a unique geographical challenge. Aviation happens in particular places by particular people. But its climate warming effects are global. This in turn makes all places less safe, whether they participate in aviation or not. In critical geography we use the term "spatial justice" to describe this challenge: how cause and effect, those who harm and those harmed, are separated by geography (Chatterton 2010). Whether your city has an airport or not, it should still take a stand on airport expansion.

What this means in practice is twofold. There needs to be an active cross-sector movement that opposes any expansion of airports and passenger numbers in or near cities. Anti-airport expansion city-based campaigns have emerged pulling together excellent legal and media work to patiently present arguments against aviation expansion (Rootes 2013). This local action needs to be coupled with lobbying for national and inter national policy to stem aviation growth, and to rapidly bring forward affordable mobility alternatives. Overnight trains and high-speed rail are key alternatives to flying. France, for example, is leading the way in banning short-haul flights where there is an existing high-speed rail route. Europe is facing a night train renaissance as countries work together to provide long-distance, transnational routes. In the short term, then, our task is a managed decline of domestic and short-haul aviation which is more readily replaced with other options, and a longer-term goal of reducing international jet travel. A frequent-flier levy needs implementing, which would also address concerns over the socially unequal spread of flying (Hopkinson & Cairns 2021). Very simply, one or two flights is enough per person per year. If you have to fly more than this, and you have a very good reason, then you will have to pay a very large excess, which would be allocated to support the kinds of alternatives stated earlier.

In the future, other options could include slower and less fuel-intensive airships and dirigible balloons, popularly known as blimps or zeppelins, and along with solar- and wind-powered sea craft these could fill a gap for international travel. These changes will have to go hand in hand with radically reduced expectations about the speed and extent of global travel and tourism. They are not about ending international travel. They allow us to rethink it through novel, slower, and also fun, alternative travel modes. Reflecting our strategy to save the city, we need to make sure we do this with social justice at its heart, make it publicly owned, accountable and accessible to all. The global pandemic did point to how we might begin to wean ourselves off our dependency on aviation. It offers lessons on how we can still maintain a globally connected world without pushing the biosphere beyond safe limits.

To support reductions in aviation, we need to build more meaningful opportunities locally for work, leisure and retail. Work-related travel was reconfigured through Covid-19 restrictions, and may never recover to pre-pandemic levels. These trends need consolidating through local employment hubs, increased digital connectivity and affordable co-working spaces. Moreover, many people escape to other countries for holidays because domestic tourism has faced decline and underinvestment. By investing in tourist destinations at home, the trend towards international holidays can be lessened. Creating more coherent train, boat and cycle infrastructures will allow more people to travel sustainably and still access large regions. Land that is currently privately owned needs shifting into public bodies and made accessible and transformed into rewilded bioregions for sustainable tourism. Many countries have beautiful but currently inaccessible places because of private ownership by landed elites, royalty, the military or global equity firms.

Our aim is not to disconnect cities. As we respond to the converging emergencies, we can build a parallel set of opportunities for jobs, connectivity and travel that do not harm people and the planet, and are socially equitable. The best that technology can offer will need to be harnessed to this task, reconceiving airports as nodes in sustainable but slower travel. This is already happening. As people become more aware of the existential crisis we face and the direct links to carbon burning, some are voluntarily choosing to fly less. The idea of "flight shame" emerged in Scandinavia and is now motivating a movement of conscious no-fliers (Irfan 2019). As it becomes socially awkward to admit to flying, group and peer pressure may start to stimulate behaviour change. The logic goes, if people I know and trust are stopping flying, then why am I still doing it? Along with financial and regulatory levers, these kinds of behavioural changes will support a reduction of mass international jet travel.

Ideas for emergency action

- City planning can be used to ban any further airport expansion.

- Broad civic coalitions can build the case against airport expansion in terms of social inequality, jobs, air quality, noise, carbon emissions, biodiversity and land use, and ultimately promote and use direct action when required.

- Regional tourism can be promoted as alternatives to air travel and incentives offered for rail travel to replace short haul flights.

- Co-working spaces with advanced digital facilities can be set up to reduce business travel.

- Institutions can offer incentives to reduce business travel, including increased payment for rail travel.

- New ordinances and regulations can be used to ban airline advertising.

- City based sport, cultural, art and music institutions, events and personalities can be used to showcase alternatives to flying, and amplify "flight shame".

  Check out: GALBA anti-airport campaign group, Leeds; Flight Free UK; short-haul flights replaced by high-speed trains, France; Sweden's Flygskam flight shame movement; Grow Heathrow and ZAD anti-airport land squats.

Acting up for nature

As we explored in Chapter 4, cities are not separate to nature. They are fully immersed in, defined by, and partly made from it. The non-human world is all around us. But given its rapid decline at the hands of the world's apex species, who will act up for nature? Who will urgently intervene on its behalf to slow its decline? This is a profound challenge to us all. As city life grows on this finite planet, in only a few decades we have exponentially encroached on the land, resources and habitats that plants and animals have evolved in for millions of years. But this is not just a danger for non-human life on earth. Humans are destroying their own life support systems. We are fundamentally intertwined and dependent on the well-being of plant and non-human animal species. These dependencies are not often talked about or acknowledged. They go far beyond our joy of experiencing and seeing the diversity of the natural world – the solace, peace and comfort we get by being in natural environments and encountering animals. There is a whole range of essential functions that the natural world offers humans, everything from direct food and air purification to soil health, crop pollination and medicines. Human life would simply not exist if the non-human world were allowed to perish.

A central part of our quest to save the city for humans, then, is to also save it for nature. It seems obvious and almost banal to state that plants and animals cannot attend meetings or make policy. But imagine if they could. What would they say and do? What would they ask of us? We have to try to understand the non-human world from their perspective at a deeper and profound level. We have to become advocates and allies in everything we do for the diverse range of living organisms, plant life, Earth systems and geophysical features that make up our planet. Let's take a look at the current state of play.

The Earth is an amazing set of self-balancing and self-regulating systems across water, carbon, soil and air, all of which support human life. While our planet is amazing, it has not always offered the perfect conditions for human life to flourish. But the past 11,000 years since the end of the last ice age has been markedly different. Known as the Holocene epoch in geological terms, stable conditions settled across the planet which have allowed human civilization to thrive and rapidly expand. The human project shifted from one of dispersed hunter-gatherers to settled agriculturalists and civilization builders. Over the millennia, the imprint of humans on the planet has been wide-ranging and over recent centuries has increased greatly. Since the industrial revolution in particular, humans began to affect the fundamental integrity and functioning of the planet.

The period since the end of the Second World War has seen the most marked change. This is the Great Acceleration: when industrialization, urbanization and militarization rapidly surged across the globe (McNeill & Engelke 2016). We have now reached a position where humans are influencing the Earth system in ways that they have never done before. For this reason, many people refer to a new era of the Anthropocene, where humans, or more accurately a small part of humanity, are undermining the safe conditions in which we thrived in the first place.

We are now starting to understand the detail of this impact. A major study led by Johan Rockstrom at the Stockholm Resilience Centre (2009) explored human impact on nine critical areas that regulate and stabilize the Earth system. They established a safe boundary for each system beyond which humans deplete them at a greater rate than the Earth can replenish and restore them. They found that human life was exceeding five of these: land-use change especially due to deforestation, climate heating from carbon emissions, biodiversity loss and extinctions, the release of chemicals, especially nitrogen and phosphorous that end up in the soil and air, and most recently what are called novel entities such as plastics and heavy metals that are being released into nature at an alarming rate. The few areas that are still in the safe zone – fresh water supplies, ocean acidification and ozone – are slowly creeping towards the unsafe boundary.

From disconnection and domination . . .

This is where cities enter the story. The industrial revolution and the enclosure of common lands pushed people into new and rapidly expanding cities from the mid-eighteenth century onwards. This was the origins of our urban industrialized world. It was deeply disruptive and violent, entwined with the emergence of capitalism and colonialism. It was in this period that so called rational Western culture became dominant, creating a divide between human culture and an inferiorized natural world as something external to be used and tamed (Plumwood 1993). By the time of the great urban explosion in the mid-twentieth century, the division between the human and natural world was deeply established.

This tragic story of nature domination can be seen in cities across the world. Many places do strive to be more sustainable through greener transport, energy and housing. But the unfortunate reality is that the current market-dominated pro-growth model of urban life in every city across the world contributes to the destruction of the natural world. While cities are acting to reduce pollution and protect nature locally, it is their contribution to exceeding the big global planetary boundaries that they also need to urgently address. In our deeply interconnected world, no city can protect itself once planetary boundaries are exceeded. The bigger parallel challenge in saving the city, then, is to connect with an agenda for global responsibility, and decouple city life from unsustainable global practices that are undermining the broader Earth system. One of the significant global challenges, for example, is to halt deforestation, especially in pristine forests where logging is being undertaken to facilitate cattle grazing or plantation crops such as coffee, nuts and oils for global markets. Global supply chains directly connect city consumption with these unsustainable global practices, and it is here that we also need to focus our efforts, through for example supporting the production of locally made alternatives.

The current experience of urban nature for many is one of disconnection. Urban sprawl, dereliction, suburban retail, highways and industry restrict urban residents' connection with the natural systems that underpin daily well-being. It is little wonder that many of us do not understand, value or want to protect the natural world. The idea of shifting baseline syndrome (Soga & Gaston 2018) helps to understand what is going on. As the natural world is increasingly eroded and damaged year by year, we forget what it was like. The increasingly urbanized and less nature-based world around us becomes the new baseline, and we shift to that as the new normal. It is a race to the bottom. Each generation has less connection and interest, and before we know it, the natural world is a historic artefact seen in museums or in wildlife documentaries, something designed for leftover spaces between steel, concrete and brick.

. . . to reconnection and stewardship

Our challenge is to comprehensively get nature back into the heart of city development, to purify air, capture carbon, clean water, generate energy, reverse species decline and support human well-being. Restorative and regenerative practices need to be central to urban policy and planning decisions (Girardet 2014). This includes approaches such as rewilding, permaculture, urban agriculture, continuous productive urban landscapes, biophilic design and blue-green infrastructure. Natural systems need to be revalued not as degradable, replaceable and free resources, but as key to climate safe and resilient cities.

A central principle is reconnection and making space for nature recovery. Cities of the future may not look anything like the cities of the past. One key feature will be expansive and connected nature corridors so wildlife can move freely over very large areas. This will involve restricting highways growth, puncturing holes through the existing road network, shrinking residential and retail suburban sprawl, creating continuous tree canopy cover and green spaces. But it is not about creating a wall between humans and non-humans. It is about creating a new relationship between the two so they can coexist, and this means valuing and protecting indigenous and regenerative farming communities throughout the world who already play a key role in stewarding the natural world. This is a huge challenge to the assumptions of the urban industrial development model that has free range over the planet's surface. These kinds of reconnection and recovery projects have already started through urban and transregional forest initiatives. In my own region, the Great North Forest, for example, spans the old industrial belt of northern England from Liverpool to Hull with an ambitious plan to plant over 50 million trees over the next 20 years. We can start to see these as bioregions: identifiable areas demarcated by natural features such as water catchments or mountain ranges in which humans can find a new role through stewardship and reciprocity rather than extraction and competition (Thackara 2017).

Reconnecting with nature can offer many advantages for humans. For example, biophilia is emerging as an urban design approach which can replicate the experiences of nature in cities in ways that promote a deeper connection between humans and non-humans (Beatley 2011). Biomimicry is also being used in cities to emulate the complex engineering and design principles found in the natural world in ways that can tackle climate breakdown, water stresses, declining air quality and biodiversity loss (Pedersen Zari & Hecht 2020). Practical applications to urban design include hybrid natural-city features such as living walls, rooftop farms, vertical gardens, water-centric design and breathing buildings. These are not just tokenistic add-ons. These features start to fundamentally change the function and form of the built world. No longer are buildings just containers for human life. They become devices to support the reconnection between humans and nature.

Saving the city, then, requires saving nature, creating a new deal between humans and our natural world not based on extractivism and commodification, but on stewardship and reconnection. This new deal for nature will support all our other moves. It will allow us to create beautiful, fulfilling places and homes that rely on and sustain carbon storage, water quality, air cleansing, biodiversity and soil health. It will inform how we move around differently beyond car dependency in ways that do not disconnect humans from humans, and humans from nature. It will show the folly of mass international travel and how to travel slower and reconnect with our wonderful world more locally. The new purpose of our thriving city economies becomes to care for and nourish each other and our natural world. That alone would consume much of the time and intelligence of human endeavour for the foreseeable future. And it is the path to our own safe and prosperous future.

Ideas for emergency action

- Urban food production can be accelerated on rooftops and by reclaiming spaces such as car parks, retail malls, golf courses and derelict development sites.

- A network of city colleges can be set up to train people in advanced horticulture, permaculture and market gardening.

- Cities can create and implement formal rewilding plans.

- The planning system can be used to create biodiversity banks, eliminate all further losses of biodiversity and habitats, and impose strict nature-based design codes on all new developments.

- All new developments need to include features to sequester carbon, harvest renewable energy, increase urban cooling and be resilient to locally relevant extreme weather events (drought, floods, heat, fire).

- The highways system can be reduced to give priority to the free flow of nature rather than vehicles.

- A network of continuous productive urban landscapes and community supported agriculture schemes can connect food producers and city consumers.

  Check out: Städte wagen Wildnis (Cities Dare Wilderness), Hannover; Curridabat's Cuidad Dulce (Sweet City) Programme, Costa Rica; Cloudburst Plan, Copenhagen; Michigan Urban Farming Initiative.

Who runs this place?

Imagine all our city-saving moves have taken place. Impressive technical and infrastructure innovations have rolled out across cities. New ways of organizing, financing and making policy are emerging. Green spaces are expanding. But it still feels like there is something missing. What has it all been for if many people do not feel part of it, empowered and involved in the unfolding story of our cities? The question is: whose zero-carbon city is it anyway? It is not enough to see climate positive changes emerge. It is not enough to sit in a café drinking a latte made using 100 per cent renewable power knowing that the people who served you are deeply unhappy, working in precarious, low-paid conditions. It is not enough to know that home insulation is being installed when there is still violence against women and people of colour, that those seeking sanctuary are rejected, or that children are still malnourished. Electric vehicles are not enough if it just consolidates the power of the world's largest vehicle manufacturers, and e-SUVs are allowed to dominate streets and kill children.

If change is to mean something, it needs to have equality and justice at its core. We need to be co-creators and co-owners of our beautiful new world. It has to change life opportunities universally, as well as regenerate the biosphere. As the inspirational anarchist and feminist Emma Goldman allegedly remarked to a guest at a party: "It's not my revolution if I can't dance to it". This is why the social emergency is such an important part of saving the city. Our efforts are aimed at more rights and recognition, but they are also a quest for freedom: not just from the drudgery of work, the mundanity of mass schooling, and from violence and oppression, but the freedom to have the ability and capacity to fulfil one's potential, without undermining the ability of others to do so. This pursuit is at the heart of our moves to save the city. Let's take a closer look at what is holding this back.

Changing city governance

First, our quest to save the city is also about changing the way city governance works. As we explored in the Introduction, the triple emergencies are complex to understand and to respond to. They are wicked problems that require new approaches to problem solving. City-level decision making often exists in topic-based siloes such as transport, health, housing, education, children's services or green spaces. On one level this is a response to making sense of a complex world and doing work in areas that people already recognize. These kinds of areas are useful for repetitive tasks where activities need to be done and the public kept safe, for example, refuse collection, building standards or child safety. But taking emergency action and making breakthroughs in the face of existential crises requires working across teams, ideas and institutions. The state has a key role in this cross-working by using its regulatory, tax raising and legal powers. But beyond that, we need a more holistic and interconnected take on key areas that support our well-being. For example, transport is actually about mobility and how we all have access to affordable, green travel options. Economic development is really about making sure we all have a thriving livelihood. Education is increasingly about practice-based transition skills for a green circular and regenerative economy. Housing is really about the wider context that homes sit within – our agenda for emergency placemaking. The bigger point is that all of these connect with each other. Budgets therefore need to be flexible and cross-departmental, and we need multi-skilled policy makers and practitioners who are trained to understand how different elements of a system shape and influence other parts. Every aspect of our life affects every other aspect, as well as others around us, and our natural world. The longer we make decisions and allocate resources in siloes, the less we can meaningfully understand and respond to a deeply interconnected world.

Second, we need to reconnect the governing authorities and institutions with the citizens they serve. There has been a significant power shift in city governance in recent years. The local state apparatus has been stripped of resources and powers, squeezed into a role of facilitating the expansion of corporate capital or providing the bare minimum of citizen services (Jessop 2002). Public assets are sold to shore up city finances as core budgets shrink. The less land and resources cities control, the less they can shape their future. This is significant to our aims as land and buildings will need to be redeployed to tackle the triple emergencies. So what kind of local state do we need? While serving the public interest is crucial, a situation where public bodies have all the power and control is not optimal. No one organization can manage the complexities of our converging emergencies. While private interests alone do not serve us well, there are also concerns over accountability and efficiency if the state becomes too big. The past few decades have seen various experiments in terms of this relationship between local state and other actors. A dominant thread has seen the state closely aligned with big business, especially through what are called public–private partnerships. While these were often well intentioned, they largely ceded too much power to private-sector business models. The delivery of basic infrastructure that underpins our daily life became privatized, extracting profit rather than ensuring safe, thriving lives (Miraftab 2004).

We need to strike a new balance. A broader move to reclaim the state is required to bring the immense powers and resources of our city authorities closer to the public they serve (Wainwright 2003). The state can play a coordinating and enabling role by using its legislative and regulatory powers to serve and protect, and also to set the vision and resources so cross-sector teams can bring forward solutions. This is a new participatory local state, a form of city governance that sets the conditions for a new civic social economy, through public–common partnerships (Milburn & Russell 2021). In these partnerships, the local state enables and empowers communities to take control and ownership of assets, and puts them in the driving seat of infrastructure renewal, popular democracy and common wealth building. City commons have a value beyond which the market can dictate. They are part of our life support system for a more equal and green future. They create a foundation that ensures our basic well-being – health, transport, energy, housing – and these should be subject to public and community control and management in perpetuity.

Third, we need to rebuild trust in local politics. The growing disenchantment and disillusionment with formal politics is at a low point across the world, and manifests itself in a range of ways from low voter turnout, mistrust of policing, the rise of populist leaders, or a retreat into social media. These shifts stem from a number of factors: lived experience of harm at the hands of authority figures; a lack of tangible progress with aspects that underpin people's well-being; the cognitive gap between what people hear about the challenges and the responses they see around them; a sense that politics is not about ordinary issues or indeed notions of a broader public good, but more about careerism, personal gain, and facilitating the needs of higher-income groups.

A new city politics

A commitment to coproduction is a key aspect of a new politics that can respond to these issues and rebuild trust. Coproduction is an approach to doing and making together, based on an ethics of collaborative and interactive design, listening and creating and testing durable solutions (Chatterton *et al.* 2018). An approach that codesigns solutions is essential given the multi-faceted and complex nature of the emergencies we face. What we are realizing is that no single organization has the insights and capacities to navigate the future. Coproduction then is a new kind of civic politics that can bring forward solutions to disrupt the status quo and meet identified challenges.

We need to include and empower a much broader cross-section of the population. But more than this, structural power and resources need devolving to groups that traditionally get marginalized, especially young people and people of colour. This is what Ben Barber (1984) called building "strong democracy", and it is a huge challenge to the way local politics is undertaken, often through negotiations with competing factions. Climate emergency declarations are instructive here. After declaring a climate emergency at a city-level, many places floundered with the huge changes needed. One of the key lessons is that the change process starts with changing the way local politics is done and how the city organizes itself and tackles problems. While much is known about the technical issues of carbon reduction at a city scale, the real challenge is outlining a new story and rationale for the city, engaging the hearts and minds of citizens, understanding, listening and empathizing with fears and concerns, gathering ideas for positive action, and getting to grips with the kinds of changes in everyday practices that need to be made within institutions and neighbourhoods. Ultimately what is needed is the co-creation of a citizen-led "Emergency Plan", along with neighbourhood-based structures which can discuss, create ownership and build consensus for the kinds of radical ideas in this book. All this has to be coupled with clear targets, milestones, resources and accountability structures.

Citizens' assemblies have emerged as a way to develop consensus and break deadlock on difficult issues (Bryant & Stone 2020). Citizens' assemblies are typically a group of people who respond to a public call and then are selected through a process of sortition, with the aim that they broadly represent the diversity of the population, to deliberate on a defined issue that needs resolving. The design of a citizens' assembly provides an opportunity for a cross-section of the public to hear from experts and campaigners over an extended period and to engage in considered, thoughtful and reasoned discussion of the issues, while calibrating action to the challenges. The end goal of the assembly is to work closely with the municipality, to make recommendations and play a scrutiny and oversight function, holding elected leaders to account on key milestones and deliverables. In my own work in Leeds, we posed the question "how can our city respond to the emergency of climate change?" As a result of deliberations over eight weeks, a jury came up with 12 ground-breaking recommendations including the public control of buses, stopping aviation expansion and a green new city deal (Leeds Climate Commission 2019). Cities across the world have undertaken citizens' assemblies on climate change, creating a new consensus on what to do. What this highlights is that members of the public, once fully informed, are prepared to go further than policy makers and politicians and provide a social licence for radical ideas (Ross *et al.* 2021).

Citizens' assemblies are essential. But our work should not end there. Standing mechanisms are required for permanent citizen deliberation as a counter to established power to propose and monitor city-based action on the social, climate and nature emergencies. More fundamentally, citizens need to be involved in crowd sourcing and co-creating the broader vision and action plan for the city. Involving as many people as possible takes time, but ultimately creates a direction of travel and set of priorities that are more durable. Policy ideas can be constantly harvested using open-source platforms, then scrutinized and filtered by panels of randomly selected citizens and policy makers. This not only creates a sense of excitement but also opens up policy making to the, as yet unknown, genius and insights of all its citizens. Finally, these policy ideas can be subject to participatory budget setting, allowing citizens to determine what precious city resources are spent on. This is a new democratic triple lock for saving our city. Public–common partnerships manage and steward land and assets. Crowd-sourced manifestos and citizens' assemblies explore what actions meet the challenges and gives a new social licence to city politicians for transformational, safe and just action. Coproduction and participatory budgets can create a longer-term social legitimacy to what actually gets done next.

Ideas for emergency action

- Citizen assemblies and juries can be used to scope and test radical policy ideas.

- Participatory budgeting can allocate a proportion of city budgets according to priorities set by communities.

- Live planning labs can be used to coproduce solutions and offer public access to real time decisions.

- Young people, women, and people of colour need to be given structural power and representation over major decisions made in the city.

- Racial and climate justice need to be given priority within city planning and policy through scrutiny and accountability committees, and reparations charters.

- Proportional forms of voting should be used for local elections, along with inclusive committee structures for decision making, led by those traditionally under-represented in decision making.

Check out: Leeds Climate Change Citizen's Jury; Lisbon's Green Participatory Budget; Better Reykjavik open platform; the Atonement and Reparation Motion, Bristol; the LEAP Manifesto, Canada.

* * *

This concludes our exploration of emergency moves that can help save the city. I invite you to develop others that are relevant to you. They will all require harnessing the power of our city players who can deploy the learn–act–build strategy. There is no guaranteed recipe for success, but together, they do increase the chances of positively and urgently addressing our converging emergencies. This book is based on a realistic hope for a better future, and that it can still turn out well. To this end, the final chapter explores what cities could look like after our decade of transformation.

6

It's the 2030s and we are saving the city

Today is my fourteenth birthday. My alarm sounds at 9 am. No more rushing for me. The new dedicated cycle lane means I can cycle to school quickly and safely. No more boring bus or wading through rush-hour traffic! And now they've opened the local job hub, my mum has a work-space down there. She doesn't miss the commute, or the cost of diesel. The hub used to be the office of one of the world's biggest law firms. But they abandoned it a few years ago when the government hiked tax for big corporations. They also introduced a law that said that no bosses could earn more than ten times their lowest paid worker. Imagine that. No more billionaires! Power to the people, my gran says.

Anyway, my dad does get up earlier than us, but he's really happy to do that now he runs the local café. All the corporate coffee houses were sold to community businesses as part of the Great Community Reform Act. It gave a huge boost to local jobs. Now loads of people round here are running their own businesses. All the adults officially work four days a week which makes them happier and have more time for other stuff and not being so grouchy. After the government got all the big firms to pay their tax there seemed to be more money around. It paid for free public transport, insulating homes, free education and healthcare and capped rent and energy bills. That was a smart move as the price of gas has gone through the roof, what with the big wars between some gas-producing nations.

Today, my mum is working down the innovation hub, where she's experimenting with rewilding the local area. Our class is visiting and helping to change what was once our local highway into what the adults call a biodiversity park. Now that there's a tram system across the whole city most of the big highways were shut. The highway used to be a nightmare. It was the scariest thing for me. Six lanes to cross, and one of my mates and his mum were badly hurt there. Now we've taken all the concrete up and we are laying plants that will help with flood control in the local river. Last week we planted some trees on what used to be the centre of the road. This ecologist we are doing it with said the plants will help keep summer temperatures down. The city introduced a "carparks for trees" programme – planting one million trees on reclaimed car parks. They are going to be busy down there.

Most people travel less these days anyway. There's loads more options for shopping and fun locally. Our favourite is the local skatepark and outdoor heated pool, made on the grounds of what was a retail mall. Imagine that. Car parks replaced by a pool! As a family, we can travel all over our region on our free public transport pass. It's helped us to do vacations differently. There's so much to do on our doorstep, especially since the regional government declared a huge part of our area a biodiversity reserve. You can cycle for miles on what used to be main roads staying in beautiful eco-lodges built on land that was animal hunting grounds. Our favourite is the guided walks with the nature rangers. There's some amazing wildlife to see now they've got more room. I never knew they existed on our doorstep. My dad is terrified. He reckons we are going to find bears, wolves and snakes. The ranger says he might soon! Flying is the one thing I miss. It's become really expensive after the frequent flier levy came in. Most airlines have pretty much stopped their flights now, and I've heard they are going to start airships which sounds cool. If you really need to get to the other side of the world, then you can use your one flight a year quota. But fast trains fill the gaps for most journeys. We get to see our cousins who live in a nearby country and we go by train. It does take a bit longer but it's a really special trip, and the trains are super cheap.

My mum says it was a clever move of the government to start taxing land not income. Now all the big landowners are selling off their land. Community businesses get first refusal, and they are building eco-housing that is affordable. My best mate Sally lives in the one next to our house. It's super interesting. They have a community factory where they built the homes using a computer machine cutter. It's based on the wiki model, where you can select and machine print your own home. They are so beautiful and we love visiting. There's loads of trees to climb, a big play area, and hang out space for the adults so they can keep out of our way!

Our house is great too. The government made a law that community cooperatives can buy empty private apartments at a huge discount. So now we pay rent to our cooperative rather than the private landlord who my mum and dad hated. They said he always put the rent up too much and didn't fix things. My dad asked him to insulate it, and he threatened to throw us out. But now the cooperative has done all that. With new heat pumps and solar panels we pay almost no energy bills. My mum wanders round in her shorts saying we are on vacation all the time! We were also given powers to close the street in front of the apartment block. We made that into a play park and community orchard. All the neighbours and local kids hang out there. It is noisier but everyone agrees it's much more interesting and safe now. And we don't need the space for so many cars anyway, as lots of people have an electric bike in the local bike hub. It's just quicker to get around on ebikes, and every house got one at half price as part of the car scrappage deal. My older brother says he misses his car, but me and my mates call him a fossil fuel dinosaur and tell him if he misses driving he can play a video game.

Things aren't perfect and there's still loads to do. People grumble about Community Fridays when everyone gets a paid day off from their normal job or school to help out locally. Some say it's enforced community service. But most people actually like what it does for us. Old folk get visited, local areas get looked after, you can help out at the local shelter, or the community farm, or people can join in at the local decision-making council. We love this one. Every month, we are all asked to make proposals for improvements and they get upvoted online. The most popular ideas are looked at by a group of people who are randomly elected to join the citizens' assembly. Then they are put forward to the city's participatory budget. My little sister was once on it as part of the youth council when they voted on making the airport into a nature and play park.

People still talk about climate change a lot. Every few months the government puts out data on how we are doing on carbon emissions. For years it has been going down. All the adults talk about the old days when we were heading for runaway global heating. Now they say we are heading for less than 2 degrees and we might make the magic target of 1.5 degrees. But there's a fair way to go. My teacher says the old guard are still clinging on to power. The big energy giants have agreed to leave the remaining oil, coal and gas in the ground. But it's still being dug up in some places, and we haven't shut down all the oil fields and coal mines yet. Who would give up that wealth without getting something in return? And there's still lies and confusion. People say that all the changes are a big conspiracy theory to take things away from ordinary people. But that doesn't make sense. Life has never been so good. We feel safer, there's more to do and people are less pressured and worried about having enough money. Basic stuff like food, transport and energy is affordable.

It's still not great for everyone. My mum says until women feel safe to walk around at night there won't be real progress, and my mate Jim is sick of being stopped by the police just because of the colour of his skin. We have been learning about climate justice at school: that the people least responsible for all the problems are dealing with the worst effects. That's just not fair. It's got to get better for everyone. There's no point us creating some kind of paradise here while the rest of the world burns! You see it on TV. We've all become aware of the damage and harm we are doing across the world. Thousands of people head towards richer countries because of food shortages and energy blackouts. Each town round here has opened up a welcome centre for refugees. We volunteer down there once a month. You meet people from all over the world, but also our own country. We met a family who lived by the coast but lost everything from flooding and had to move into temporary accommodation. Our apartment block has taken in a family for a while. They left their country when their farm collapsed in a drought. They are such brave people and are desperate to go home. But it's great having them around. We hang out with their kids. They say when they can return home, we can come and stay with them whenever we want.

I've got my birthday party this afternoon in the common room of our apartment bock. Then we are going to the outdoor city pool and skatepark on our bikes. I feel hopeful about the future. I always wonder what my life will be like by 2040 when I'm an adult. I see my mum and dad happier too. For the first time, they seem optimistic about how everything will turn out. It feels like our city has turned a corner. The new mayor is super popular. She keeps bringing in new reforms to make life better. Her next big things are a three-day working week, free local renewable energy for every house, and a guaranteed wage for everyone. My dad says we are saving the city. I said to him, how do you save a city? Well, he says, that's a longer story . . . .

Online resources to support emergency action

Climate and social movements

Black Lives Matter   https://blacklivesmatter.com/
Extinction Rebellion   https://rebellion.global/
Fridays for the Future   https://fridaysforfuture.org/
Just Stop Oil   https://juststopoil.org/
Reclaim the Power   https://reclaimthepower.org.uk
Youth Strike for Climate   https://ukscn.org/about-us/

Climate emergency

Anti Aviation group   https://stay-grounded.org/about/
C40 Cities   https://www.c40.org/
Centre for Alternative Technology   https://cat.org.uk/
Climate Action Network   https://climatenetwork.org/
Climate Emergency Centres   https://climateemergencycentre.co.uk/
Fossil Fuel Divestment   https://gofossilfree.org/
IPCC   https://www.ipcc.ch/
Rapid Transition Alliance   https://www.rapidtransition.org/

Nature-based solutions

Biophilic Cities   https://www.biophiliccities.org/
CPULs   https://blogs.brighton.ac.uk/pulr/cpul-concept/
Half Earth Project   https://www.half-earthproject.org

Permaculture Association   https://www.permaculture.org.uk/
Rewilding   https://www.rewildingbritain.org.uk/

Community and economy solutions

15-minute neighbourhoods   https://www.15minutecity.com/

B Corps   https://www.bcorporation.net/en-us/

Centre for Story based Strategies   https://www.storybasedstrategy.org/

Circular Economy   https://ellenmacarthurfoundation.org/topics/
circular-economy-introduction/overview

Citizen Assemblies   https://citizensassembly.co.uk/

Cohousing Network   https://cohousing.org.uk/

Community Land Trusts   https://www.communitylandtrusts.org.uk/

Community Wealth Building   https://cles.org.uk/
the-community-wealth-building-centre-of-excellence/

Decolonising Economics   https://decolonisingeconomics.org/

Doughnut Economics Action Lab   https://doughnuteconomics.org/

Four Day Week   https://www.4dayweek.co.uk/

Green New Deal   https://www.greennewdealuk.org/

Housing Cooperatives   https://www.communityledhomes.org.uk/
what-housing-co-operative

Just Transitions   https://climatejusticealliance.org/just-transition/

Participatory Budgeting   https://pbnetwork.org.uk/

Self Help Housing   https://www.communityledhomes.org.uk/
what-self-help-housing

Universal Basic Services   https://universalbasicservices.org/

Universal Basic Income   https://www.ubilabnetwork.org/

Urban Labs   https://www.ucl.ac.uk/urban-lab/about/partnerships/urban-lab

References

Alexander, S. & B. Gleeson 2020. "Urban social movements and the degrowth transition: towards a grassroots theory of change". *Journal of Australian Political Economy* 86: 355–78.

Alinsky, S. 1971. *Rules for Radicals.* New York: Vintage.

Art of Hosting 2011. Workbook. https://storage.ning.com/topology/rest/1.0/file/get/2836583049?profile=original.

Anderson, B. 2017. "Cultural geography 1: intensities and forms of power". *Progress in Human Geography* 41(4): 501–11.

Angelis, M. De 2003. "Reflections on alternatives, commons and communities". *The Commoner* 6(Winter): 1–14.

Appadurai, A. 2001. "Deep democracy: urban governmentality and the horizon of politics". *Environment & Urbanization* 13(2): 23–43.

Ayni Institute 2021. "Ayni Institute Web Site". https://ayni.institute/socialchange/.

Barber, A. 2021. *Consumed: The Need for Collective Change – Colonialism, Climate Change & Consumerism.* London: Brazen.

Barber, B. 1984. *Strong Democracy: Participatory Politics for a New Age.* Oakland, CA: University of California Press.

Barnett, J. 2016. *City Design: Modernist, Traditional, Green and Systems Perspectives.* 2nd ed. New York: Routledge.

Bärnthaler, R., A. Novy & L. Plank 2021. "The foundational economy as a cornerstone for a social-ecological transformation". *Sustainability* 13: 10460.

Barrett, J. *et al.* 2013. "Consumption-based GHG emission accounting: a UK case study". *Climate Policy* 13(4): 451–70.

Barry, J. 2019. "Planning in and for a post-growth and post-carbon economy". In S. Davoudi *et al.* (eds), *The Routledge Companion to Environmental Planning.* Abingdon: Routledge.

Bassens, D., W. Kębłowski & D. Lambert 2020. "Placing cities in the circular economy: neoliberal urbanism or spaces of socio-ecological transition?" *Urban Geography* 41(6): 893–7.

Bauen, A. *et al.* 2020. "Sustainable aviation fuels: status, challenges and prospects of drop-in liquid fuels, hydrogen and electrification in aviation". *Johnson Matthey Technology Review* 64(3): 263–78.

Beard, V. & D. Mitlin 2021. "Water access in global South cities: the challenges of intermittency and affordability". *World Development* 47(2): 105625.

Beatley, T. 2011. *Biophilic Cities: Integrating Nature into Urban Design and Planning*. Washington, DC: Island Press.

Benavav, A. 2022. *Automation and the Future of Work*. London: Verso.

Berglund, O. & D. Schmidt 2020. *Extinction Rebellion and Climate Change Activism: Breaking the Law to Change the World*. Cham, CH: Palgrave Macmillan.

Bernays, E. 1928. *Propaganda*. New York: Horace Liveright.

Bloomberg, M. & C. Pope 2017. *Climate of Hope*. New York: St Martin's Press.

Bock, L. & U. Burkhardt 2019. "Contrail cirrus radiative forcing for future air traffic". *Atmospheric Chemistry and Physics* 19(12): 8163–74.

Bookchin, M. 2006. *Social Ecology and Communalism*. Chico, CA: AK Press.

Bookchin, M. & D. Foreman 1991. *Defending the Earth: A Debate*. Montreal: Black Rose Books.

Bowman, A. & J. Froud (eds) 2014. *The End of the Experiment? From Competition to the Foundational Economy*. Manchester: Manchester University Press.

Brenner, N. (ed.) 2015. *Implosions/Explosions. Implosions/Explosions*. Berlin: JOVIS Verlag.

Brenner, N. & N. Theodore 2002. "Cities and the geographies of 'actually existing neoliberalism'". *Antipode* 34(3): 349–79.

Bridle, J. 2018. *New Dark Age: Technology and the End of the Future*. London: Verso.

Bryant, P. & L. Stone 2020. "Climate assemblies and juries: a people powered response to the climate emergency". A report By Shared Future. https://sharedfuturecic.org.uk/wp-content/uploads/2020/08/Shared-Future-PCAN-Climate-Assemblies-and-Juries-web.pdf.

Bulkeley, H. & P. Newell 2015. *Governing Climate Change*. Abingdon: Routledge.

Burgmann, V. 2002. *Power, Profit and Protest: Australian Social Movements and Globalisation*. London: Routledge.

Burgueño Salas, E. 2022. "Total fuel consumption of commercial airlines worldwide between 2005 and 2022". Statista Web Site. https://www.statista.com/statistics/655057/fuel-consumption-of-airlines-worldwide/.

Butcher, M. & K. Maclean 2018. "Gendering the city: the lived experience of transforming cities, urban cultures and spaces of belonging". *Gender, Place & Culture* 25(5): 686–94.

Carter, A. & J. McKenzie 2020. "Amplifying 'keep it in the ground' first-movers: toward a comparative framework". *Society & Natural Resources* 33(11): 1339–58.

Castree, N. & M. Sparke 2000. "Introduction: professional geography and the corporatization of the university: experiences, evaluations, and engagements". *Antipode* 32(3): 222–9.

Cavalieri, P. 2001. *The Animal Question: Why Nonhuman Animals Deserve Human Rights.* Oxford: Oxford University Press.

Center for Story-based Strategy 2020. "What is story-based strategy". https://www.storybasedstrategy.org/what-is-storybased-strategy.

Centre for Local Economic Strategies 2020. "Owning the economy: community wealth building 2020". Manchester.

Chambers, J. *et al.* 2021. "Six modes of co-production for sustainability". *Nature Sustainability* 4 (Nov): 983–96.

Chancel, L. 2020. *Unsustainable Inequalities: Social Justice and the Environment.* Cambridge, MA: Harvard University Press.

Chandler, D. & J. Reid 2016. *The Neoliberal Subject: Resilience, Adaptation and Vulnerability.* London: Rowman & Littlefield.

Chatterton, P. 2010. "Seeking the urban common: furthering the debate on spatial justice". *City: Analysis of Urban Change, Theory, Action* 14(6): 625–8.

Chatterton, P. 2015. *Low Impact Living: A Field Guide to Ecological, Affordable Community Building.* London: Routledge.

Chatterton, P. 2017. "Being a Zapatista wherever you are: reflections on academic-activist practice from Latin America to the UK". In P. North & M. Scott Cato (eds), *Towards Just and Sustainable Economies: The Social and Solidarity Economy North and South*, 235–56. Bristol: Policy Press.

Chatterton, P. 2020. "Coronavirus: we're in a real-time laboratory of a more sustainable urban future". The Conversation UK, April.

Chatterton, P. *et al.* 2018. "Recasting urban governance through Leeds City Lab: developing alternatives to neoliberal urban austerity in co-production laboratories". *International Journal of Urban and Regional Research* 42(2): 226–43.

Chatterton, P. & A. Pusey 2020. "Beyond capitalist enclosure, commodification and alienation: postcapitalist praxis as commons, social production and useful doing". *Progress in Human Geography* 44(1): 27–48.

Childs, M. & P. de Zylva 2021. "A dangerous distraction: why offsetting will worsen the climate and nature emergencies". A report by Friends of the Earth. https://policy.friendsoftheearth.uk/sites/default/files/documents/2021-10/Dangerous_distractions_report_October_2021.pdf.

Christophers, B. 2022. *Rentier Capitalism: Who Owns the Economy, and Who Pays for It?* London: Verso.

Civil Contingencies Act 2004. UK Government.

Clarke, N. & A. Cochrane 2013. "Geographies and politics of localism: the localism of the United Kingdom's coalition government". *Political Geography* 34: 10–23.

Coaffee, J. & P. Lee 2016. *Urban Resilience: Planning for Risk, Crisis and Uncertainty*. London: Palgrave Macmillan.

Connell, R., B. Fawcett & G. Meagher 2009. "Neoliberalism, new public management and the human service professions: introduction to the special issue". *Journal of Sociology* 45(4): 331–8.

Coote, A. & A. Percy 2020. *The Case for Universal Basic Services*. Cambridge: Polity.

Cowen, D. 2020. "Following the infrastructures of empire: notes on cities, settler colonialism, and method". *Urban Geography* 41(4): 469–86.

Cowtan, G. 2017. *Community Energy: A Guide to Community-Based Renewable-Energy Projects*. Cambridge: Green Books.

Cumbers, A. 2016. "Remunicipalization, the low-carbon transition, and energy democracy". In *State of the World* edited by Worldwatch Institute. Washington, DC: Island Press.

Darling, J. 2010. "A city of sanctuary: the relational re-imagining of Sheffield's asylum politics". *Transactions of the Institute of British Geographers* NS 35: 125–40.

DeNardis, L. 2020. *The Internet in Everything: Freedom and Security in a World with No Off Switch*. New Haven, CT: Yale University Press.

Dhillon, J. (ed.) 2022. *Indigenous Resurgence: Decolonialization and Movements for Environmental Justice*. New York: Berghahn.

Dinerstein, A. *et al.* 2019. "Scaling up or deepening? Developing the radical potential of the SSE sector in a time of crisis UK". Draft paper prepared in response to the UNTFSSE call for papers 2018 Implementing the Sustainable Development Goals: What Role for Social and Solidarity Economy?

Doughnut Economics Action Lab 2021. "Turning the ideas of doughnut economics into action". https://doughnuteconomics.org/.

Edwards, M. & D. Leonard 2022. "Effects of large vehicles on pedestrian and pedalcyclist injury severity". *Journal of Safety Research* 82: 275–82.

Eisenstein, C. 2018. *Climate: A New Story*. Berkeley, CA: North Atlantic Books.

Emery, T. & J. Thrift 2021. "20-minute neighbourhoods: creating healthier, active, prosperous communities – an introduction for council planners in England". A report by the Town and Country Planning Association. https://tcpa.org.uk/wp-content/uploads/2021/11/final_20mnguide-compressed.pdf.

European Court of Auditors 2019. "Audit preview: roads connecting European regions". Luxembourg. https://www.eca.europa.eu/mt/publications?did=49952.

Evans, J. & A. Karvonen 2014. "'Give me a laboratory and I will lower your carbon footprint!' Urban laboratories and the governance of low-carbon futures". *International Journal of Urban and Regional Research* 38(2): 413–30.

Evans, S. 2021. "Analysis: which countries are historically responsible for climate change?" Carbon Brief, October 2021.

Eviction Lab 2022. "Eviction filings by city, USA". https://evictionlab.org/eviction-tracking/.

Ewert, A. & Y. Chang 2018. "Levels of nature and stress response". *Behavioral Sciences* 8: 49.

Extinction Rebellion 2021. "Extinction Rebellion: tell the truth". https://extinctionrebellion.uk/the-truth/.

Farías, I. & T. Bender (eds) 2011. *Urban Assemblages: How Actor-Network Theory Changes Urban Studies*. London: Routledge.

Fath, B. *et al.* 2019. "Measuring regenerative economics: 10 principles and measures undergirding systemic economic health". *Global Transitions* 1: 15–27.

Featherstone, D. 2012. *Solidarity: Hidden Histories and Geographies of Internationalism*. London: Zed Books.

Fennell, P. *et al.* 2022. "Going net zero for cement and steel". *Nature* 603: 574–7.

Floater, G. *et al.* 2016. "Co-benefits of urban climate action: a framework for cities". September. London: Economics of Green Cities Programme.

Foxon, T. 2011. "A coevolutionary framework for analysing a transition to a sustainable low carbon economy". *Ecological Economics* 70(12): 2258–67.

Forsberg, M. & C. Bleil de Souza 2021. "Implementing regenerative standards in politically green Nordic social welfare states: can Sweden adopt the living building challenge?" *Sustainability* 13(2): 738.

Florida, R. 2004 *The Rise of the Creative Class*. London: Routledge.

Friends of the Earth 2021. "Aviation and climate change: our position". London.

Fry, T. 2017. *Remaking Cities: An Introduction to Urban Metrofitting*. London: Bloomsbury.

Fuller, D. 2008. "Public geographies: taking stock". *Progress in Human Geography* 32(6): 834–44.

Gaventa, J. 2021. "Linking the prepositions: using power analysis to inform strategies for social action". *Journal of Political Power* 14(1): 109–30.

Geels, F. 2010. "Ontologies, socio-technical transitions (to sustainability), and the multi-level perspective". *Research Policy* 39: 495–510.

Gehl, J. 2008. *Life Between Buildings: Using Public Space*. Copenhagen: Danish Architectural Press.

Ghaemi, Z. & A. Smith 2020. "A review on the quantification of life cycle greenhouse gas emissions at urban scale". *Journal of Cleaner Production* 252 (119634).

Gibson-Graham, J. 2008. "Diverse economies: performative practices for 'other worlds'". *Progress in Human Geography* 32(5): 613–32.

Gibson-Graham, J., J. Cameron & S. Healy 2013. *Take Back the Economy: An Ethical Guide for Transforming Our Communities*. Minneapolis, MN: University of Minnesota Press.

Girardet, H. 2014. *Creating Regenerative Cities*. London: Routledge.

Glaeser, E. & D. Cutler 2021. *Survival of the City: Living and Thriving in an Age of Isolation*. New York: Penguin.

Gollan, D. 2022. "Private aviation is booming and bursting at the seams". *Forbes*, 23 May.

Goodway, D. 2012. *Anarchist Seeds beneath the Snow: Left-Libertarian Thought and British Writers from William Morris to Colin Ward*. Liverpool: Liverpool University Press.

Gopnik, A. 2016. *The Gardener and the Carpenter*. New York: Farrar, Straus & Giroux.

Gore, T. 2020. "Confronting carbon inequality: putting climate justice at the heart of the COVID-19 recovery". Oxfam Media Briefing.

Gorz, A. 1973. "The social ideology of the motorcar". *Le Sauvage*, Sep–Oct. http://unevenearth.org/2018/08/the-social-ideology-of-the-motorcar/.

Gough, I. 2019. "Universal basic services: a theoretical and moral framework". *Political Quarterly* 90(3): 534–42.

Gould, K. & T. Lewis 2012. "The environmental injustice of green gentrification: the case of Brooklyn's Prospect Park". In J. DeSena & T. Shortell (eds), *The World in Brooklyn: Gentrification, Immigration and Ethnic Politics in a Global City*. Lanham, MD: Lexington Books.

Gouverneur, D. 2015. *Planning and Design for Future Informal Settlements: Shaping the Self-Constructed City*. Abingdon: Routledge.

Guoping, L. & H. Zhou 2015. "Globalization of financial capitalism and its impact on financial sovereignty". *World Review of Political Economy* 6(2):176–91.

Graeber, D. 2019. *Bullshit Jobs: The Rise of Pointless Work, and What We Can Do About It*. London: Penguin.

Graham, S. 2010. *Cities Under Siege: The New Military Urbanism*. London: Verso.

Granovetter, M. 1973. "The strength of weak ties". *American Journal of Sociology* 78(6): 1360–80.

Green, B. 2020. *The Smart Enough City: Putting Technology in Its Place to Reclaim Our Urban Future*. Cambridge, MA: MIT Press.

Gruszka, K. 2017. "Framing the collaborative economy: voices of contestation". *Environmental Innovation and Societal Transitions* 23: 92–104.

Gunder, F. 1966. "The development of underdevelopment". *Monthly Review* 18(4).

Gudde, P. *et al.* 2021. "The role of UK local government in delivering on net zero carbon commitments: you've declared a climate emergency, so what's the plan?" *Energy Policy* 154(112245).

Hagedorn, G. *et al.* 2019. "Concerns of young protesters are justified | Letters". *Science* 364(6436): 139–40.

Haines, G. 2022. "The best countries for social progress". Positive New. https://www.positive.news/society/ranked-the-best-countries-for-social-progress/.

Hall, M., D. Scott & S. Gössling 2013. "The primacy of climate change for sustainable international tourism". *Sustainable Development* 21: 112–21.

Hardt, M. & A. Negri 2004 *Multitude: War and Democracy in an Age of Empire*. London: Penguin.

Harvey, D. 2013. *Rebel Cities: From the Right to the City to the Urban Revolution*. London: Verso.

Harvey. D. 2017. *Marx, Capital and the Madness of Economic Reason*. Oxford: Oxford University Press.

Hawken, P. (ed.) 2018. *Drawdown: The Most Comprehensive Plan Ever Proposed to Reverse Global Warming*. London: Penguin.

Hayes, N. 2021. *The Book of Trespass: Crossing the Lines That Divide Us*. London: Bloomsbury.

Milburn, K. & B. Russell 2021. "Public–common partnerships: democratising ownership and urban development". A report by Common-Wealth. https://www.common-wealth.co.uk/publications/public-common-partnerships-building-new-circuits-of-collective-ownership.

Hess, C. & E. Ostrom (eds) 2006. *Understanding Knowledge as a Commons: From Theory to Practice*. Cambridge, MA: MIT Press.

Hickel, J. 2021. *Less Is More: How Degrowth Will Save the World*. London: Penguin.

Hoggan, J. & R. Littlemore 2009. *Climate Cover-Up: The Crusade to Deny Global Warming*. Vancouver, BC: Greystone Books.

Hollands, R. 2008. "Will the real smart city please stand up? Intelligent, progressive or entrepreneurial?" *City* 12(3): 303–20.

Holloway, J. 2010. *Crack Capitalism*. London: Pluto.

Holloway, J. 2022. *Hope in Hopeless Times*. London: Pluto.

Homer-Dixon, T. 2006. *The Upside of Down: Catastrophe, Creativity and the Renewal of Civilisation*. London: Souvenir Press.

hooks, bell. 2003. *Teaching Community: A Pedagogy of Hope*. London: Routledge.

Hopkins, P. 2018. "Feminist geographies and intersectionality". *Gender, Place and Culture* 25(4): 585–90.

Hopkinson, L. & S. Cairns 2021. "Elite status: global inequalities in glying". A report by Possible. https://www.wearepossible.org/latest-news/elite-status-how-a-small-minority-around-the-world-take-an-unfair-share-of-flights.

Hopkinson, L. & L. Sloman 2018. "More than electric cars. Why we need to reduce traffic to reach carbon targets". Friends of the Earth Briefing. https://www.transportforqualityoflife.com/u/files/1%20More%20than%20electric%20cars%20briefing.pdf.

Horton, M. & P. Freire 1990. *We Make the Road by Walking: Conversations on Education and Social Change*. Philadelphia, PA: Temple University Press.

Hulme, M. 2019. "Climate emergency politics is dangerous". *Issues in Science and Technology* 36(1): 23–5.

ICAO 2019. "The world of air transport in 2019". ICAO Annual Report. United Nations International Civic Aviation Organisation.

IEEP & Oxfam 2021. "Carbon inequality in 2030: per capita consumption emissions and the 1.5°C goal". Oxford.

Illich, I. 1971. *Deschooling Society*. Harmondsworth: Penguin.

Incropera, F. 2015. *Climate Change: A Wicked Problem: Complexity and Uncertainty at the Intersection of Science, Economics, Politics, and Human Behavior*. New York: Cambridge University Press.

International Labour Organization 2021. "The future of work in the automotive industry". Note on the proceedings, Geneva, 15–19 February.

IPBES 2019. *Global Assessment Report on Biodiversity and Ecosystem Services of the Intergovernmental Science-Policy Platform on Biodiversity and Ecosystem Services*. Edited by E. Brondizio *et al.* Bonn: IPBES Secretariat.

IPCC (Intergovernmental Panel on Climate Change) 2018. *Global Warming of 1.5°C. An IPCC Special Report on the Impacts of Global Warming of 1.5°C above Pre-Industrial Levels and Related Global Greenhouse Gas Emission Pathways, in the Context of Strengthening the Global Response to the Threat of Climate Change, Sustainable Development, and Efforts to Eradicate Poverty*. Edited by V. Masson-Delmotte *et al.* IPCC.

IPCC 2021. *Climate Change 2021: The Physical Science Basis. Contribution of Working Group I to the Sixth Assessment Report of the Intergovernmental Panel on Climate Change*. Edited by V. Masson-Delmotte *et al.* Cambridge: Cambridge University Press.

IPCC 2022. *Climate Change 2022: Impacts, Adaptation and Vulnerability. Working Group II Contribution to the IPCC Sixth Assessment Report*. Edited by H.-O. Pörtner *et al.* Cambridge: Cambridge University Press.

Irfan, U. 2019. "Air travel is a huge contributor to climate change: a new global movement wants you to be ashamed to fly". Vox, November.

Jafry, T. (ed.) 2019. *Routledge Handbook of Climate Justice*. Abingdon: Routledge.

Jakob, M. & J. Hilaire 2015. "Unburnable fossil-fuel reserves". *Nature* 517: 150–52.

Jensen, O. 2006. "'Facework', flow and the city: Simmel, Goffman, and mobility in the contemporary city". *Mobilities* 1(2): 143–65.

Jessop, B. 2002. "Liberalism, neoliberalism, and urban governance: a state-theoretical perspective". *Antipode* 34(3): 452–72.

Kärcher, B. 2018. "Formation and radiative forcing of contrail cirrus". *Nature Communications* 9: 1824.

Keil, R. 2010. "The urban politics of roll-with-it neoliberalization". *City* 13(2/3): 230–45.

Kempin Reuter, T. 2019. "Human rights and the city: including marginalized communities in urban development and smart cities". *Journal of Human Rights* 18(4): 382–402.

Knight Frank 2021 *The Wealth Report 2021*. London: Knight Frank Research.

Kropotkin, P. 1974. *Fields, Factories and Workshops Tomorrow*. London: Freedom Press.

Kühne, K. *et al.* 2022. "'Carbon bombs': mapping key fossil fuel projects". *Energy Policy* 166(112950).

Kwarteng, K. *et al.* 2012. "Britannia unchained". In *Britannia Unchained*, 110–12. London: Palgrave Macmillan.

Lang, R., P. Chatterton & D. Mullins 2020. "Grassroots innovations in community-led housing in England: the role and evolution of intermediaries". *International Journal of Urban Sustainable Development* 12(1): 52–72.

Lansley, S. & H. Reed 2019. *Basic Income for All: From Desirability to Feasibility*. London: Compass.

Lees, L., H. Shin & E. López-Morales 2016. *Planetary Gentrification*. Cambridge: Polity.

Leeds Climate Commission 2019. Leeds citizens' jury recommendations. https://www.leedsclimate.org.uk/leeds-citizens-jury-recommendations-published.

Leeds Doughnut Coalition 2021. "The first Leeds Doughnut City Portrait: towards a safe and thriving city for all". https://www.climateactionleeds.org.uk/_files/ugd/6c95b1_9d4dec776a724feabdc443a84527c45f.pdf.

Lehmann, S. (ed.) 2015. *Low Carbon Cities: Transforming Urban Systems*. Abingdon: Routledge.

Lent, J. 2021. *The Web of Meaning: Integrating Science and Traditional Wisdom to Find Our Place in the Universe*. London: Profile.

Lewis, P. & R. Evans 2013. *Undercover: The True Story of Britain's Secret Police*. London: Faber & Faber.

Li, Q. 2010. "Effect of forest bathing trips on human immune function". *Environmental Health and Preventive Medicine* 15: 9–17.

Linebaugh, P. 2014. *Stop, Thief! The Commons, Enclosures, and Resistance*. Oakland, CA: PM Press.

Lloyd Goodwin, T. & H. Power 2021. "Community wealth building: a history". A report by the Centre for Local Economic Strategies. https://cles.org.uk/publications/community-wealth-building-a-history/.

MacKinnon, D. & K. Derickson 2012. "From resilience to resourcefulness: a critique of resilience policy and activism". *Progress in Human Geography* 37(2): 253–70.

Macy, J. & C. Johnstone 2012. *Active Hope: How to Face the Mess We're in Without Going Crazy*. Novato, CA: New World Library.

Madden, D. 2019. "Editorial: city of emergency". *City* 23(3): 281–4.

Maeckelbergh, M. 2011. "Doing is believing: prefiguration as strategic practice in the alterglobalization movement". *Social Movement Studies* 10(1): 1–20.

Malm, A. 2021. *How to Blow Up a Pipeline*. London: Verso.

Marquis, C. 2021. *Better Business: How the B Corp Movement Is Remaking Capitalism*. New Haven, CT: Yale University Press.

Massey, D. 2004. "Geographies of responsibility". *Geografiska Annaler: Series B, Human Geography* 86: 5–18.

Mathiesen, K. 2015. "How and where did UK lose city-sized area of green space in just six years?" *The Guardian*, 2 July. https://www.theguardian.com/environment/2015/jul/02/how-where-did-uk-lose-green-space-bigger-than-a-city-six-years.

Mayer, M. 2009. "The 'right to the city' in the context of shifting mottos of urban social movements". *City* 13(2/3): 362–74.

McArdle, R. 2021. "Intersectional climate urbanism: towards the inclusion of marginalised voices". *Geoforum* 126 (Aug): 302–05.

McNeill, J. & P. Engelke 2016. *The Great Acceleration: An Environmental History of the Anthropocene Since 1945*. Cambridge, MA: Harvard University Press.

Meadows, D. 1999. "Leverage points: places to intervene in a system". The Sustainability Institute. https://donellameadows.org/wp-content/userfiles/Leverage_Points.pdf.

Meadows, D. *et al.* 1972. *Limits to Growth*. Washington, DC: Other Potomac Associates.

Mertes, T. (ed.) 2003. *A Movement of Movements: Is Another World Really Possible?* London: Verso.

Meyer, A. 2004. "Briefing: contraction and convergence". *Proceedings of the Institution of Civil Engineers: Engineering Sustainability* 157(ES4): 189–92.

Midnight Notes 1990. *New Enclosures*. Los Angeles, CA: Semiotext(e).

Mignolo, W. & C. Walsh 2018. *On Decoloniality: Concepts, Analytics, Praxis*. Durham, NC: Duke University Press.

Minow, M. 2021. "Equality vs. equity". *American Journal of Law and Equality* 1: 167–93.

Miraftab, F. 2004. "Public–private partnerships: the Trojan horse of neoliberal development?" *Journal of Planning Education and Research* 24: 89–101.

Mitchell, K. (ed.) 2008. *Practising Public Scholarship: Experiences and Possibilities Beyond the Academy*. Chichester: Wiley-Blackwell.

Braungart, M., W. McDonough & A. Bollinger 2007. "Cradle-to-cradle design: creating healthy emissions: a strategy for eco-effective product and system design". *Journal of Cleaner Production* 15(13/14): 1337–48.

Molotch, H. 1976. "The city as a growth machine: toward a political economy of place". *American Journal of Sociology* 82(2): 309–32.

Monbiot, G. 2001. *Captive State: The Corporate Takeover of Britain*. London: Pan.

Monbiot, G. 2009. "Activists like the Drax protesters are the conscience of the nation". *The Guardian*, 3 July.

Monbiot, George *et al.* 2019. "Land for the many: changing the way our fundamental asset is used". A report to the UK Labour Party. https://labour.org.uk/wp-content/uploads/2019/06/12081_19-Land-for-the-Many.pdf.

Mora, L. *et al.* (eds) 2022. *Sustainable Smart City Transitions: Theoretical Foundations, Sociotechnical Assemblage and Governance Mechanisms*. Abingdon: Routledge.

Moyer, B. *et al.* 2001. "Doing democracy: the MAP model for organizing social movements". *Journal of Family Social Work* 8.

Mueller, G. 2021. *Breaking Things at Work: The Luddites Are Right About Why You Hate Your Job*. London: Verso.

McDonald, R. *et al.* 2020. "Research gaps in knowledge of the impact of urban growth on biodiversity". *Nature Sustainability* 3: 16–24.

Nakate, V. 2021. *A Bigger Picture: My Fight to Bring a New African Voice to the Climate Crisis*. London: Pan.

North, P. & M. Scott Cato (eds) 2017. *Towards Just and Sustainable Economies: The Social and Solidarity Economy North and South*. Bristol: Policy Press.

Nussbaum, M. 2000. *Women and Human Development: The Capabilities Approach*. Cambridge: Cambridge University Press.

O'Connor, C. & J. Weatherall 2020. *The Misinformation Age: How False Beliefs Spread*. New Haven, CT: Yale University Press.

Ocasio-Cortez, A. 2019. "The green new deal". AOC Campaign. https://www.ocasiocortez.com/green-new-deal.

Our World in Data 2022. "Access to electricity, urban". https://ourworldindata.org/grapher/access-to-electricity-urban-vs-rural?tab=table.

Papa, E. 2020. "Would you ditch your car if public transport was free? Here's what researchers have found". The Conversation UK, March.

Partzsch, L. 2017. "'Power with' and 'power to' in environmental politics and the transition to sustainability". *Environmental Politics* 26(2): 193–211.

Paskaleva, K., J. Evans & K. Watson 2021. "Co-producing smart cities: a quadruple helix approach to assessment". *European Urban and Regional Studies* 28(4): 395–412.

Peeters, P. *et al.* 2016. "Are technology myths stalling aviation climate policy?" *Transportation Research Part D* 44: 30–42.

Perelman, M. 2000. *The Invention of Capitalism: Classical Political Economy and the Secret History of Primitive Accumulation.* Durham, NC: Duke University Press.

Pettifor, A. 2019. *The Case for the Green New Deal.* London: Verso.

Plumwood, V. 1993. *Feminism and the Mastery of Nature.* London: Routledge.

Polanyi, K. 1944. *The Great Transformation: The Political and Economic Origins of Our Time.* New York: Farrar & Rinehart.

Purcell, M. 2006. "Urban democracy and the local trap". *Urban Studies* 43(11): 1921–41.

Pusey, A. 2017. "Towards a university of the common: reimagining the university in order to abolish it with the really open university". *Open Library of Humanities* 3(2): 1–27.

Rakodi, C. & T. Lloyd-Jones (eds) 2002. *Urban Livelihoods: A People-Centred Approach to Reducing Poverty.* London: Routledge.

Raworth, K. 2017. *Doughnut Economics: Seven Ways to Think like a 21st Century Economist.* London: Penguin.

Recio, E. & D. Hestad 2022. "Indigenous peoples: defending an environment for all". International Institute for Sustainable Development. https://www.iisd. org/articles/deep-dive/indigenous-peoples-defending-environment-all.

Rifkin, J. 2013. *Third Industrial Revolution: How Lateral Power Is Transforming Energy.* London: Palgrave Macmillan.

Ritchie, H. 2020a. "Cars, planes, trains: where do CO2 emissions from transport come from?" Our World in Data. https://ourworldindata.org/ co2-emissions-from-transport.

Ritchie, H. 2020b. "Climate change and flying: what share of global CO_2 emissions come from aviation?" Our World in Data. https:// ourworldindata.org/co2-emissions-from-aviation.

Robertson, B. 2015. *Holacracy: The Revolutionary Management System That Abolishes Hierarchy.* New York: Henry Holt.

Robinson, M. 2018. *Climate Justice: Hope, Resilience and the Fight for a Sustainable Future.* London: Bloomsbury.

Rockström, J. *et al.* 2009. "Planetary boundaries: exploring the safe operating space for humanity". *Ecology and Society* 14(2): 32

Rootes, C. 2013. "From local conflict to national issue: when and how environmental campaigns succeed in transcending the local". *Environmental Politics* 22(1): 95–114.

Rosenberg, M. 2015. *Nonviolent Communication: A Language of Life*. Encinita, CA: PuddleDancer Press.

Ross, A. *et al.* 2021. "Deliberative democracy and environmental justice: evaluating the role of citizens' juries in urban climate governance". *Local Environment* 26(12): 1512–31.

Routledge, P. & A. Cumbers 2009. *Global Justice Networks: Geographies of Transnational Solidarity*. New York: Palgrave Macmillan.

Russell, B. 2019. "Beyond the local trap: new municipalism and the rise of the fearless cities". *Antipode* 51(3): 989–1010.

Saad, L. 2020. *Me and White Supremacy*. London: Quercus.

Sahan, E. *et al.* 2022. "What Doughnut Economics means for business". Doughnut Economics Action Lab. https://doughnuteconomics.org/tools/191.

Satterthwaite, D. *et al.* 2020. "Building resilience to climate change in informal settlements". *One Earth* 2(2): 143–56.

Saunter, N. 2022. "The Thriving Places Index: a positive way to measure human flourishing". https://www.rapidtransition.org/commentaries/the-thriving-places-index-a-positive-way-to-measure-human-flourishing/.

Seyfang, G. & A. Smith 2007. "Grassroots innovations for sustainable development: towards a new research and policy agenda". *Environmental Politics* 16(4): 584–603.

Scholz, T. & N. Schneider 2016. *Ours to Hack and Own: The Rise of Platform Cooperatives*. London: OR Books.

Sheila McKechnie Foundation 2019. "Social power: the social change grid". London.

Shiva, V. 2020. *Reclaiming the Commons: Biodiversity, Traditional Knowledge, and the Rights of Mother Earth*. Santa Fe, NM: Synergetic Press.

Shrubsole, G. 2020. *Who Owns England? How We Lost Our Land and How to Take It Back*. London: HarperCollins.

Silva, B., M. Khan & K. Han 2018. "Towards sustainable smart cities: a review of trends, architectures, components, and open challenges in smart cities". *Sustainable Cities and Society* 38: 697–713.

Simpson, J. 2017. "Finding brand success in the digital world". *Forbes*, 25 August.

Sims, R. *et al.* 2014. "Transport". In *Climate Change 2014: Mitigation of Climate Change. Contribution of Working Group III to the Fifth Assessment Report of the Intergovernmental Panel on Climate Change*, edited by O. Edenhofer *et al.* Cambridge: Cambridge University Press.

Smith, A., A. Stirling & F. Berkhout 2005. "The governance of sustainable socio-technical transitions". *Research Policy* 34: 1491–510.

Smith, N. 2006. "There's no such thing as a natural disaster". *SSRC Items: Insights from the Social Sciences*, June.

Soga, M. & K. Gaston 2018. "Shifting baseline syndrome: causes, consequences, and implications". *Frontiers in Ecology and the Environment* 16(4): 222–30.

Sparke, M. 2008. "Political geography – political geographies of globalization III: resistance". *Progress in Human Geography* 32(3): 423–40.

Stainforth, T. & B. Brzezinski 2020. "More than half of all CO2 emissions since 1751 emitted in the last 30 years". Institute for European Environmental Policy, April.

Statista 2020. "Distribution of oil demand in the OECD in 2020, by sector". https://www.statista.com/statistics/307194/top-oil-consuming-sectors-worldwide/.

Stokstad, E. 2020. "The pandemic stilled human activity: what did this 'anthropause' mean for wildlife?" *Science* 13 (Aug), online.

Stronge, W. & A. Harper 2019. "The shorter working week: a radical and pragmatic proposal". A report by Autonomy. https://autonomy.work/portfolio/the-shorter-working-week-a-report-from-autonomy-in-collaboration-with-members-of-the-4-day-week-campaign/.

Sundberg, J. 2007. "Reconfiguring North-South solidarity: critical reflections on experiences of transnational resistance". *Antipode* 39(1): 144–66.

Taylor, M. 2018. "London mayor unveils plan to tackle 'climate emergency'". *The Guardian*, 11 December. https://www.theguardian.com/uk-news/2018/dec/11/london-mayor-sadiq-khan-city-climate-emergency.

Taylor, M. 2020. "The evolution of Extinction Rebellion". *The Guardian*, 4 August.

Thackara, J. 2017. "Back to the land: design agenda for bioregions". https://thackara.com/urbanrural/back-to-the-land-2-0-a-design-agenda-for-bioregions/.

Thaler, R. & C. Sunstein 2008. *Nudge: Improving Decisions About Health, Wealth and Happiness*. New Haven, CT: Yale University Press.

Thunberg, G. 2019a. "'Our house is on fire': Greta Thunberg, 16, urges leaders to act on climate". *The Guardian*, 25 January.

Thunberg, G. 2019b. "Greta Thunberg speech: I have a dream that the powerful will take the climate crisis seriously". *The Independent*, 20 September.

Toffler, A. 1980. *The Third Wave*. New York: Bantam.

Trapese Collective (ed.) 2007. *Do It Yourself: A Handbook for Changing Our World*. London: Pluto.

Tulder, R. van & N. Keen 2018. "Capturing collaborative challenges: designing complexity-sensitive theories of change for cross-sector partnerships". *Journal of Business Ethics* 150(2): 315–32.

UN 2022. "UN Sustainable Development Goals". United Nations Department of Economic and Social Affairs. https://sdgs.un.org/goals.

UN Development Group 2017. "Theory of change: UNDAF companion guidance". Vol. June.

University of Oxford 2020. "What is net zero?" https://netzeroclimate.org/.

Vacca, J. 2020. *Solving Urban Infrastructure Problems Using Smart City Technologies: Handbook on Planning, Design, Development, and Regulation*. Amsterdam: Elsevier.

Valero, A. *et al.* 2021. "Are 'green' jobs good jobs? How lessons from the experience to-date can inform labour market transitions of the future". London School of Economics. https://www.lse.ac.uk/granthaminstitute/wp-content/uploads/2021/10/Are-Green-Jobs-Good-Jobs_Full-report-4.pdf.

Vogel, J. *et al.* 2021. "Socio-economic conditions for satisfying human needs at low energy use: an international analysis of social provisioning". *Global Environmental Change* 69: 102287.

Wa Thiong'o, N. 1992. *Decolonising the Mind: The Politics of Language in African Literature*. Nairobi: East African Publishers.

Wainwright, H. 2003. *Reclaim the State: Experiments in Popular Democracy*. London: Verso.

Wainwright, H. & D. Elliott 1981. *Lucas Plan: New Trade Unionism in the Making*. London: Allison & Busby.

Waite, D. & K. Morgan 2019. "City deals in the polycentric state: the spaces and politics of metrophilia in the UK". *European Urban and Regional Studies* 26(4): 382–99.

Wallace-Wells, D. 2019. *The Uninhabitable Earth: Life After Warming*. New York: Tim Duggan Books.

Wallace-Wells, D. 2020. "Climate change and the future of humanity". RSA Minimate. London: RSA.

Ward, B. & J. Lewis 2002. "Plugging the leaks: making the most of every pound that enters your local economy". New Economics Foundation. https://neweconomics.org/uploads/files/plugging-the-leaks.pdf.

Ward, C. & A. Fyson 1973. *Streetwork: The Exploding School*. London: Routledge & Kegan Paul.

Welsby, D. *et al.* 2021. "Unextractable fossil fuels in a 1.5°C world". *Nature* 597 (9 Sep): 230–40.

WHO 2005. "Bridging the 'know–do' gap: meeting on knowledge translation in global health". Geneva: World Health Organization.

Williams, J. 2021. *Climate Change Is Racist: Race, Privilege and the Struggle for Climate Justice*. London: Icon.

Wilson, D. & A. Jonas (eds) 1999. *The Urban Growth Machine: Critical Perspectives, Two Decades Later*. Albany, NY: SUNY Press.

Wilson, E. 2016. *Half-Earth: Our Planet's Fight for Life*. New York: Liveright.

Wollen P. & J. Kerr 2003. *Autopia: Cars and Culture*. London: Reaktion.

World Bank and Climate Watch 2020. "CO2 emissions (metric tons per capita)". https://data.worldbank.org/indicator/EN.ATM.CO2E.PC? locations=US&most_recent_value_desc=false.

World Economic Forum 2021. "Clean skies for tomorrow leaders: 10% sustainable aviation fuel by 2030". https://www.weforum.org/press/2021/ 09/clean-skies-for-tomorrow-leaders-commit-to-10-sustainable-aviation-fuel-by-2030/.

Wright, E. 2010. *Envisioning Real Utopias*. London: Verso.

Zari, M. P. 2018. *Regenerative Urban Design and Ecosystem Biomimicry*. Abingdon: Routledge.

Zari, M. P. & K. Hecht 2020. "Biomimicry for regenerative built environments: mapping design strategies for producing ecosystem services". *Biomimetics* 5(2): 18.

Zibechi, R. 2012. *Territories in Resistance: A Cartography of Latin American Social Movements*. Oakland, CA: AK Press.

Zhen, W. *et al.* 2019. "Changing urban green spaces in Shanghai: trends, drivers and policy implications". *Land Use Policy* 87: 104080.

Index